Workbook to Accompany
Masonry Skills

Workbook to Accompany
Masonry Skills
Sixth Edition

Stephen Richard

THOMSON

DELMAR LEARNING Australia Brazil Canada Mexico Singapore Spain United Kingdom United States

Masonry Skills, Sixth Edition
Stephen Richard

Vice President, Technology and Trades ABU:
David Garza

Director of Learning Solutions:
Sandy Clark

Managing Editor:
Larry Main

Senior Acquisitions Editor:
James DeVoe

Senior Product Manager:
John Fisher

Marketing Director:
Deborah S. Yarnell

Marketing Manager:
Kevin Rivenberg

Marketing Specialist:
Mark Pierro

Director of Production:
Patty Stephan

Production Manager:
Stacy Masucci

Content Project Manager:
Jennifer Hanley

Editorial Assistant:
Tom Best

Cover Image:
GettyImages, Inc./Medioimages/Photodisc collection

COPYRIGHT © 2008 Thomson Delmar Learning, a division of Thomson Learning Inc. All rights reserved. The Thomson Learning Inc. logo is a registered trademark used herein under license.

Printed in the United States of America
1 2 3 4 5 XX 09 08 07

For more information contact Thomson Delmar Learning
Executive Woods
5 Maxwell Drive, PO Box 8007,
Clifton Park, NY 12065-8007
Or find us on the World Wide Web at
www.delmarlearning.com

ALL RIGHTS RESERVED. No part of this work covered by the copyright hereon may be reproduced in any form or by any means—graphic, electronic, or mechanical, including photocopying, recording, taping, Web distribution, or information storage and retrieval systems—without the written permission of the publisher.

For permission to use material from the text or product, contact us by
Tel.   (800) 730-2214
Fax   (800) 730-2215
www.thomsonrights.com

Library of Congress Cataloging-in-Publication Data:

ISBN-10: 1-4180-3757-5
ISBN-13: 978-1-4180-3757-4

NOTICE TO THE READER

Publisher does not warrant or guarantee any of the products described herein or perform any independent analysis in connection with any of the product information contained herein. Publisher does not assume, and expressly disclaims, any obligation to obtain and include information other than that provided to it by the manufacturer.

The reader is expressly warned to consider and adopt all safety precautions that might be indicated by the activities herein and to avoid all potential hazards. By following the instructions contained herein, the reader willingly assumes all risks in connection with such instructions.

The publisher makes no representation or warranties of any kind, including but not limited to, the warranties of fitness for particular purpose or merchantability, nor are any such representations implied with respect to the material set forth herein, and the publisher takes no responsibility with respect to such material. The publisher shall not be liable for any special, consequential, or exemplary damages resulting, in whole or part, from the readers' use of, or reliance upon, this material.

Table of Contents

Preface vii

Introduction viii

SECTION 1 SAFETY PRACTICES, TOOLS, EQUIPMENT, AND BASIC TOOL SKILLS

STUDENT COMPETENCY #1	Describe and list personal protective equipment for a masonry worker	3
STUDENT COMPETENCY #2	Stocking the masonry project	5
STUDENT COMPETENCY #3	Spreading mortar and buttering head joints for brick	7
STUDENT COMPETENCY #4	Spreading mortar to lay block	11
STUDENT COMPETENCY #5	Cutting masonry units with the brick hammer and brick set chisel	15

SECTION 2 DEVELOPMENTS AND MANUFACTURE OF BRICK AND CONCRETE MASONRY UNITS

STUDENT COMPETENCY #6	Identifying and explaining the common ways to use ten masonry units	21

SECTION 3 LAYING BRICK TO A LINE AND BUILDING A BRICK CORNER

STUDENT COMPETENCY #7	Mixing mortar	25
STUDENT COMPETENCY #8	Laying brick to the line with corner poles or speed leads	29
STUDENT COMPETENCY #9	Laying brick to the line and building a masonry opening using corner poles or speed leads	33
STUDENT COMPETENCY #10	Laying brick veneer on running bond and a soldier course to the line with corner poles or speed leads	37
STUDENT COMPETENCY #11	Building a brick rack-back lead	41
STUDENT COMPETENCY #12	Building a brick corner	45
STUDENT COMPETENCY #13	Laying brick corners and filling in the wall	49

SECTION 4 MORTAR AND ESSENTIALS OF BONDING

STUDENT COMPETENCY #14	Laying brick to the line with quoin corners using corner poles or speed leads	55
STUDENT COMPETENCY #15	Laying brick corners and filling in the wall with common bond	59
STUDENT COMPETENCY #16	Laying brick corners and filling in the wall with English bond	63
STUDENT COMPETENCY #17	Laying brick corners and filling in the wall with Flemish bond	67
STUDENT COMPETENCY #18	Laying a 4" block pier with corbelled brick top	71
STUDENT COMPETENCY #19	Laying brick corbelling and coping	75

SECTION 5 LAYING CONCRETE BLOCK

STUDENT COMPETENCY #20	Laying block to the line	81
STUDENT COMPETENCY #21	Laying block to the line and building in a masonry opening and rowlock sill	85
STUDENT COMPETENCY #22	Laying 4″ × 8″ × 16″ split face block veneer to the line	89
STUDENT COMPETENCY #23	Laying block to the line and building a masonry opening spanned by precast concrete lintels	93
STUDENT COMPETENCY #24	Laying an 8″ corner lead	97
STUDENT COMPETENCY #25	Laying a 10″ corner lead	103
STUDENT COMPETENCY #26	Laying a 12′ corner lead	107
STUDENT COMPETENCY #27	Laying a 6″ corner lead	111
STUDENT COMPETENCY #28	Laying a 4″ corner lead	115
STUDENT COMPETENCY #29	Building a block pier	119

SECTION 6 ESTIMATING BRICK AND CONCRETE BLOCK BY RULE OF THUMB MATH AND CUTTING WITH THE MASONRY SAW

STUDENT COMPETENCY #30	Cutting masonry units with the saw	125

SECTION 7 MASONRY PRACTICES AND DETAILS OF CONSTRUCTION

STUDENT COMPETENCY #31	Building a brick pier	133
STUDENT COMPETENCY #32	Building a six-corner brick pier	137
STUDENT COMPETENCY #33	Building a composite wall	141
STUDENT COMPETENCY #34	Brick on horizontal relief angle	145
STUDENT COMPETENCY #35	Building a cavity wall	149
STUDENT COMPETENCY #36	Building movement joints and bond beam in a composite wall	153
STUDENT COMPETENCY #37	Building pilasters and a masonry opening in a block wall	157
STUDENT COMPETENCY #38	Laying block and glass block to the line	161
STUDENT COMPETENCY #39	Laying 4″ block and glass block to the line	165

SECTION 8 SCAFFOLDING AND CLEANING MASONRY WORK

STUDENT COMPETENCY #40	Building scaffold	171

SECTION 9 CHIMNEYS AND FIREPLACES

STUDENT COMPETENCY #41	Laying a brick chimney	177
STUDENT COMPETENCY #42	Building a Rumford fireplace	181

SECTION 10 ARCHES

STUDENT COMPETENCY #43	Laying an arch in brick veneer	189
STUDENT COMPETENCY #44	Laying out and constructing a two rowlock semicircular arch	193

SECTION 11 CONCRETE AND STEPS

STUDENT COMPETENCY #45	Using concrete hand tools	199
STUDENT COMPETENCY #46	Laying brick pavers in sand	203
STUDENT COMPETENCY #47	Mixing concrete and pouring step treads	207
STUDENT COMPETENCY #48	Laying brick steps with precast concrete treads	211
STUDENT COMPETENCY #49	Laying brick steps with rowlock treads	215

Preface

This competency workbook was developed as a companion to the *Masonry Skills,* sixth edition textbook by Richard Kreh. The competency projects have gradable performance objectives for the student to achieve and the instructor to evaluate.

The competency projects are designed to measure the skills of beginning to advanced masonry students. The student will find some tips and reminders from the text within the competency-related learning activities and procedures. However this workbook does not replace the full body of masonry knowledge in the textbook.

The brick projects are mostly veneer or single wythe projects, which is the predominant way bricks are used today. Various patterns, quoin corners, and corbelling projects are included so the student can see and learn the extensive design possibilities of single wythe brick veneer.

The projects with masonry openings often have the opening offset to help the student as they learn to lay out and bond the masonry units to the openings specified in the project.

Competency #6 is to identify commonly used masonry units and how they are used. The instructor should use the type of units common to your geographical area.

Competency #42 "Building a Rumford fireplace" is based on the extensive research and development done by Jim Buckley and Dana Martini of Superior Clay and Bob Rucker of CMS Industries. This simple to build yet efficient fireplace design will give the student a challenge to build and the knowledge that they can build an all-masonry fireplace that is energy efficient and clean burning. In my thirty years of experience, this is the best fireplace design ever developed. Many other masons concur with my observations.

The student or instructor may not use all the competencies, but there are enough included to provide for training our future masonry workforce.

Introduction

A PERSONAL NOTE TO THE STUDENT

The craft of masonry requires workers with expert skills to build with the most durable, safest, and beautiful building materials the world has ever used. For our civilization to prosper in the future, beginning masons need to learn to perform high-quality masonry work in an efficient manner.

Skilled masons are needed to build durable, energy-efficient buildings and to maintain and restore the masonry treasures of the past. As you use this workbook to successfully complete each project, remember the following guidelines, because they will be the procedures on which your instructor will grade you and they will ensure your success in the future workforce.

- *Set up:*
  - Be aware of the safety guidelines of your school, training facility, or jobsite.
  - Work safely to protect yourself and others around you.
  - Properly stock your project.
  - Understand and properly lay out the project before laying units.
- *Lay up:* Build and maintain *Plumb, Level, Range,* and *Bond*[*] in the project. Build leads and set corner poles using these guidelines:

    Stay $1/16''$ away from the line when laying to the line. Understand masonry units have slight differences in dimension, but if you lay to the line, the wall will be plumb, level, and within range. Because you have accurately built the lead or set the corner pole, you should not use your level to maintain plumb, level, and range in a wall you are laying to the line.
- *Clean up:* Keep your project or work area free of trash and clutter that can cause a hazard. Clean your area and put away your tools at the end of each workday. After your project is complete, clean the area before you ask your instructor for an evaluation.

As you work at completing projects in this workbook, you may encounter situations in which you feel you need help. Before you go to your instructor, use the preceding guidelines to evaluate your project layout, masonry units, or wall—you may answer your own questions. Best wishes and safety as you learn the craft of masonry.

GRADING THE WORKBOOK PROJECTS

Each project in the workbook has an instructor evaluation chart. Each gradable phase of the project is listed and the way it will be graded. Remember proper setup and safety procedures are checked and graded on each and every project.

When evaluating for *Plumb, Level, Height, Bond,* and *Measurement,* a metal step gauge is used (available from Bon Tool, or a machine shop can make some). The back flat side of the gauge is inserted between the four-foot level and the project or the tape measure when the instructor checks the various measurements. The space created by deviation from plumb, level, height, bond, and measurement is measured by which step of the gauge fits that space. These numbers are recorded on your grade sheet. The total points recorded for that project are deducted from 100. This will give you your final grade for the competency project.

---
[*]See "How a Mason Checks the Work" section in this Introduction.

Solid steel step gauge.

The step gauge is laid flat on the top block to check for level. The level bubble is level and the first step of the gauge fits under the level; one point is deducted from the project.

SAMPLE INSTRUCTOR'S EVALUATION AND CHECKLIST[†]

	Possible Points	Points Off
1. Project set up and properly stocked. *This shall include initial layout measurement.*	15	
2. Safety procedures followed. *This includes personal protective equipment and job site safety.*	15	
3. Plumb. *Check places indicated on project plan with step gauge.*	10	
4. Level. *Check places indicated on project plan with step gauge.*	5	
5. Height. *Check places indicated on project plan with step gauge. Begin measurement from the top of first course because some floors are not level.*	10	
6. Bond. *Check for proper bond pattern. Check for places indicated on project plan with step gauge. Plumb bond of head joints may be with in a $1/4"$ plus or minus tolerance except for stack bond.*	5	
7. Tooling: *Deduct a point for each of the following:* Holes Smears (Over 1" square) Tags or "fingernails"("fingernails" are sag or slump lines at the intersection of the head and bed joints that need to be tooled) that protrude from wall Uniform shape	10	
8. Proper use of trowel and tools. *This includes proper spreading of joints and buttering head joints; properly using line and line blocks, twigs, and pins; properly using levels for measurement of plumb, level, and range; and proper use of hammer and other masonry tools.*	10	
9. Motion study as described in project procedure and good masonry practices. *Follow directions and work from a properly stocked project area with a minimum of steps. Tear down and remove project supplies and equipment in an efficient way.*	10	
10. Neatness. *Proper uniform or clothing is required. Remove all litter and masonry debris and clean up floor in project area as you are working and at the end of each work session. Clean and put away tools at the end of each workday or completion of project.*	10	
	100	

TOTAL

FINAL GRADE

[†]See Section 5, Student Competency #29

Final Grade for the Competency Project

The instructor should establish the percentage point at which the student needs to pass the competency project based on learned skill levels.

- Competency #1 and Competency #2: The student should have a 100 to pass and move on to other projects.
- Competencies #3–49:
  - Beginning students should have a 60–80% to pass (instructor's discretion).
  - Intermediate students should have 80–90% to pass (instructor's discretion).
  - Advanced students should have 90% to pass (instructor's discretion)

The instructor may adjust the pass/fail grade threshold as needed. The scoring system with each competency can still be used.

Instructor's Comments

It is helpful to the student and instructor to record skills that need improving or improvements achieved. These notes may be referred to later by the student to improve skills or by the instructor to check progress.

HOW A MASON CHECKS THE WORK

A mason may work on a project as small as a mailbox pier; as functional as a warm, dry house foundation; as big as a multistory hospital or academic building; or as complex as a high temperature industrial kiln. In each project the mason is constantly checking and evaluating *Plumb, Level, Bond, Square,* and *Range*. These concepts become second nature to a mason after years of experience. Let's briefly evaluate each one.

Plumb

- Is each unit in the wall plumb?
- Is the wall plumb even though some masonry units are not manufactured to an exact flat and square dimension?
- Are corners, jambs, and various brick positions and shapes plumb in their position in the wall?
- Is the backup structure for your masonry work plumb?
- Are anchor bolts plumb?

Plumbing a corner with a plumb rule.

Level

- Is each unit in the wall level?
- Is the wall level even though the starting point on footings and floors may be out of level?
- Is the wall level even though some masonry units are not manufactured to an exact flat and square dimension?
- Are the block and brick courses at each end of the project being built at the same level?
- To what height or "level" should the project be built?
- At what position or "level" do masonry openings get built?
- Are bearing plates level and set at the right position or "level" in the project?

Laying the first course of bricks in mortar. The mason is leveling the corner brick and the end brick before moving bricks in the center. This is the procedure to follow when working on a base that is not level.

Bond

- Which bond pattern is used on the project?
- Is there an even spacing of masonry units "bond" from one end of the project to the other?
- What masonry unit positions or mechanical ties are needed to bond or structurally tie wythes of masonry to the backup or to themselves?
- Is plumb bond (the proper alignment of head joints) being maintained?
- Is the mortar mixed and used in the right consistency to achieve proper adhesive bond?
- Have the masonry units been laid properly and efficiently the first time so the bond is not broken by readjusting units?

Brick and block composite wall tied together with metal wire joint reinforcement.

Notice how mortar does not stick properly to brick after initial set has taken place. This brick must be relaid in fresh mortar to obtain a good mortar bond.

Dry bonding the proper number of bricks for a corner nine courses high. Notice that there are five bricks leading in one direction and four in the other direction.

Square

- Are the corner locations and dimensions true to achieve a square project?
- Is the masonry opening square by maintaining plumb jambs?
- Are walls or piers square with each other in a related project?

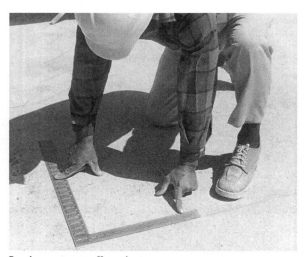
Laying out a small project.

Range

- Is the tail of each lead straightedged or "ranged" with a tight unobstructed line to the opposite corner?
- Will the face of the masonry wall remain plumb by not pushing the line or allowing units to slump because of unsuitable mortar, wet units, or workmanship?
- Is the color "range" correct by properly stocking and or mixing masonry units?

USING CORNER POLES AND SPEED LEADS

Beginning masonry students greatly benefit from "laying to the line" before they proceed to "level work" such as building corners and other projects with the level. Many of the competency projects in this workbook are developed for use with corner poles and or speed leads. Unit 10 of the *Masonry Skills*

Diagonals are checked to ensure the building is square.

Use a tight unobstructed line to "range" the tail of the lead to the far corner.

textbook shows project 4A, which is a plywood backup for a veneer wall with the ability to attach manufactured or shop-made corner poles. The student can then hang the line, lay out coursing, and lay brick or block.

Another alternative is a "speed lead," which would simply be a straight board attached to a doorjamb, end of a wall, or the face of a wall. If space allows, permanent masonry piers or corner leads can be built for the students to hang a mason's line on.

Masons laying brick with a corner pole as a guide. Notice the line attached to the pole at the top of the bricks. The wall is being laid in the running bond.

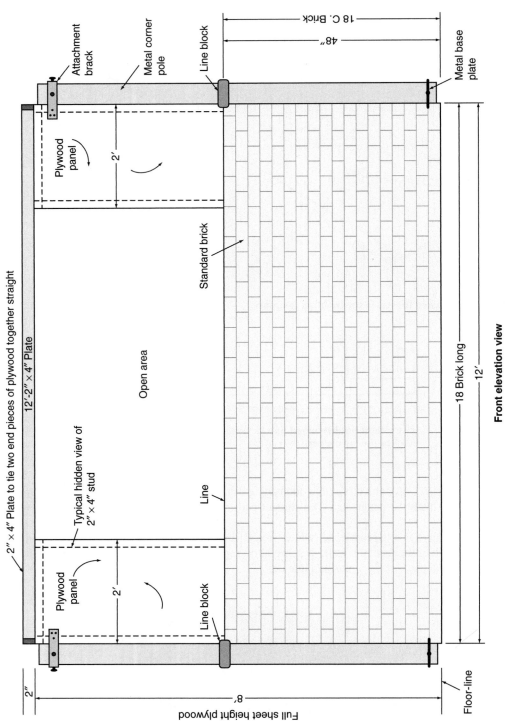

Front elevation view

Plywood frame and manufactured corner poles from Unit 10 of the *Masonry Skills* textbook.

Workbook to Accompany
Masonry Skills

SECTION ONE
SAFETY PRACTICES, TOOLS, EQUIPMENT, AND BASIC TOOL SKILLS

Student Competency #1

TASK

Describe and list personal protective equipment for a masonry worker.

PERFORMANCE OBJECTIVES

- The student will be able to list and describe the proper personal protective equipment worn by masonry workers.
- The student shall explain what job site conditions or hazards exist on a masonry job site or training facility that personal protective equipment can protect the worker from.

RELATED LEARNING ACTIVITIES

1. Read or review Unit 1 of the *Masonry Skills* textbook.
2. Visit a masonry job site or training facility.
3. Look through your safety equipment or those provided for you at your training facility.

PROCEDURE

1. Using an OSHA-approved hard hat, show the two parts of the headgear and explain how they work as a system. Explain how much clearance there should be between helmet and liner. Explain why it should never be worn backwards on the head. Describe two hazards a hard hat protects the worker from.
2. Using an approved pair of safety glasses, explain why they are used and the purpose of the side shields. Explain when it is best to use face-fitting goggles. Describe two hazards safety glasses or goggles protect the worker from.
3. Using approved earplugs or earmuffs, show how they should properly fit to protect the worker's ears. Describe two noise dangers that require hearing protection.
4. Using a dust mask or respirator, explain how they should fit to the face. Explain the difference between a dust mask and a respirator. Describe two airborne hazards on masonry job sites that require a dust mask or respirator.
5. Using proper work boots, explain what height they should be on the ankle. Describe two job site hazards that require leather work boots to provide protection.
6. Explain the purpose of steel-toed work boots. List two potential hazards that steel-toed work boots provide protection from.
7. List five potential hazards a mason with safety awareness will watch for. List hazards that could be caused by vehicles, equipment, working on scaffold or ladders, storage or stocking of materials, and extreme hot or cold weather.
8. Explain the proper procedure if there is an accident at your school shop, training facility, or job site.

Student Competency #1

Instructor's Evaluation and Checklist

Student's Name: _____

	Possible Points	Points Off

1. Student
    a. has approved hard hat.
    b. explained shell, lining, and clearance.
    c. explained two dangers.
	12	

2. Student
    a. has approved safety glasses or goggles.
    b. describes side shields and face-fitting goggles.
    c. explains two hazards.
	12	

3. Student
    a. has approved ear protection.
    b. explains proper fit of earplugs and earmuffs.
    c. describes two dangers to ears on job site.
	12	

4. Student
    a. has approved dust mask or respirator.
    b. describes proper fit of both devices.
    c. describes two hazards that require breathing protection.
	12	

5. Student
    a. has proper footwear.
    b. describes two job site hazards to the feet.
	12	

6. Student
    a. has or describes steel-toed work boots.
    b. describes two potential foot-crushing hazards.
	12	

7. List five potential hazards.
    1.
    2.
    3.
    4.
    5.
	12	

8. First response to accident or injury.	16	
	100	

TOTAL

FINAL GRADE

_____ SCORE
(100% score required to pass)

Instructor's Comments:

Student Competency #2

TASK

Stocking the masonry project.

PERFORMANCE OBJECTIVE

- The student shall properly and safely stock mortar, materials, tools, and any other related equipment to properly and efficiently build the project.

RELATED LEARNING ACTIVITIES

1. Review Unit 1 of the *Masonry Skills* textbook.
2. Learn what your training center uses to elevate mud boards or materials.
3. A properly stocked work area or wall helps the mason work quickly and efficiently.

PROCEDURE

1. Place mortarboard in center of project wall or corner area on a stand. The front edge of the mortarboard should be 2' away from the near side of the wall. If you are building a corner lead, mud board should be inside corner or outside on the corner, depending on which way you have been instructed to build the work. Either way, have the board about 2' from the near side of wall.
2. Block project: Stock the block on each side of the board with the end facing the wall and keeping the block about 2' back from near side of wall. Keep the stacks close together so they help support each other. Do not stock more than five high. When lifting and carrying block, be sure to lift properly, as explained in your *Masonry Skills* textbook.
3. For a brick project, set the brick on a block. This will eliminate excessive bending over. For brick projects above 4' high, use two 8' scaffold plank and mud stands or block and make a platform two planks wide and 2' back from face of wall. Place mud stand on the plank in center of work area and stock brick on each side.
4. Place your levels in an upright position in a block set on the floor close to your work area so you can easily get them when needed. Get the tools you will be using out of your bag or toolbox immediately. Set the rest of your tools under the mud stand or plank so they are not in the way.
5. When your project is properly stocked, ask your instructor for an evaluation. After the instructor's evaluation, you are ready to build a project.

A properly stocked project. Each item and tool is in place so the mason does not have to take extra steps.

A properly stocked project to build a corner lead.

Instructor's Evaluation and Checklist

Student's Name: _____

	Possible Points	Points Off

1. Mud board is
   a. elevated and centered.
   b. 2' from near side of wall.
	
	20
	20	

2. Block is right side up with ends facing wall and/or brick neatly stacked so they will not shift and fall over during work.	20	
3. Materials are 2' from near side of wall.	20	
4. Levels are standing up in a block and tool bags/boxes stored under mud board or material plank.	20	
	100	

TOTAL

FINAL GRADE

_____ **SCORE**
(100% score required to pass)

Instructor's Comments:

Student Competency #3

TASK

Spreading mortar and buttering head joints for brick.

PERFORMANCE OBJECTIVES

- The student shall properly spread the bed joint and butter the head joints to lay six brick.
- The student shall plumb, level, and straightedge the six brick.

RELATED LEARNING ACTIVITIES

1. Read and review Unit 3 of the *Masonry Skills* textbook.
2. Watch the demonstration by the instructor.
3. Properly stock your project according to Competency #2, "Stocking a masonry project."
4. Observe all safety procedures!
5. Learning good trowel skills is important for efficient quality masonry work.

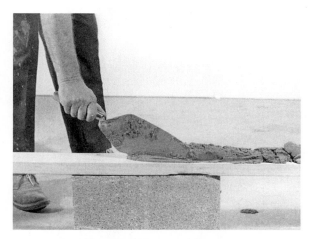

Spreading the mortar on the wall. Note the location of the trowel in the center of the board and the flat back side of the trowel turned away from the mason.

Using the trowel to cut off the excess mortar. Note the flat angle at which the trowel is held. The mason can catch the mortar so it does not fall on the floor.

8 Student Competency #3

(a)

(b)

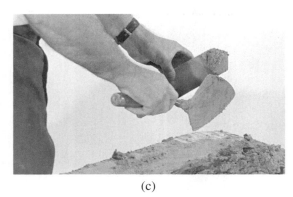
(c)

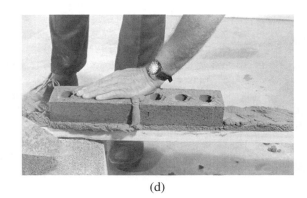
(d)

The proper way to form a head joint.

Leveling bricks by tapping them down with the blade of the trowel. Note the correct position of the trowel, parallel with the course of bricks and back from the edge of the bricks.

Plumbing the end bricks with the level.

To align the front faces of the remaining bricks, the mason places the level horizontally along the front upper edges of the bricks and taps the bricks until they are aligned. The level must be kept even with the top of each end brick while the mason adjusts the middle bricks.

PROCEDURE

1. Stand two 8″ block on end with 3′ between them. Allow 2′ between your mortarboard and these blocks. Set a 2″ × 4″ × 4′ piece of lumber on the blocks to make a raised level area to spread bed joints for bricklaying. Using the method taught by your instructor and Unit 2 in the *Masonry Skills* textbook, practice spreading a uniform bed joint on the 2″ × 4″. Be in control of your trowel so you don't drop too much mortar and can achieve an evenly furrowed bed joint.
2. When you can spread the bed joint in four swipes of the trowel and achieve a uniform bed joint of the proper thickness on the 2″ × 4″, you can start laying bricks.
3. There are two methods to lay brick.
   Method #1. Spread the bed joint and lay the first brick at the left end of the 2″ × 4″. Butter a head joint on that brick and lay the next brick to it. As you lay the brick into position, use your trowel to cut excess mud that squeezes out and return it to the board or use it to butter on your next head joint. If you are right-handed, you will be facing and moving your body to the right side of the project as you lay the brick. If you are left-handed, you will face and move your body to the left. Continue this method until all bricks are laid.
   Method #2. Spread the bed joint and lay the first brick at the right end of the 2″ × 4″. Pick up a brick and butter the head joint on the end of the brick you will lay against the brick already laid. As you lay the brick into position, use your trowel to cut excess mud as it squeezes out. If you are right-handed, you will be facing the right side of the project and backing up as you lay brick from the right side of the project to the left side. If you are left-handed, you will be facing the left side of the project and backing up as you lay brick from the left side of the project to the right side. Continue laying brick with this method until you have laid all your brick.
4. After six brick are laid, plumb each end brick even with the face of the 2″ × 4″ and straightedge the brick between with your level. Next use your four-foot level to level the top of the brick. Lightly tap the brick down to position if they are high. You may lightly tap your level with your trowel handle to uniformly level the six brick. After leveling the brick, recheck plumb and straightedge on the face of the brick.
5. Tool the brick carefully without smearing.
6. When the project is done, clean your area and ask the instructor for an evaluation.

10   Student Competency #3

Instructor's Evaluation and Checklist

Student's Name: _____

	Possible Points	Points Off
1. Project properly set up and stocked. Levels in block and tool bag put under mud board or material so it will not be tripped over. Care of tools.	10	
2. Safety procedures followed	10	
3. Bricks plumb	10	
4. Bricks level	10	
5. Six brick are straight on face or "ranged."	10	
6. Uniform 3/8" head joints	10	
7. Height. #6 modular or #5 brick spacing	10	
8. Tooling: Holes Smears Tags or "fingernails" protruding from face of wall	10	
9. Proper use of trowel	10	
10. Motion study as described in procedure and neatness. Bricks laid with face towards student.	10	
	100	

TOTAL

FINAL GRADE

_____ **SCORE**

(Required to pass: Beginning 60–80%, Intermediate 80–90%, Advanced 90%)

Instructor's Comments:

Student Competency #4

TASK

Spreading mortar to lay block.

PERFORMANCE OBJECTIVES

- The student shall be able to spread bed joints and butter head joints to lay block.
- The student shall be able to lay, plumb, level, and straightedge (range) a three-course-high rack-back lead.

RELATED LEARNING ACTIVITIES

1. Read and review Units 2 and 3 of the *Masonry Skills* textbook.
2. Watch the demonstration by your instructor.
3. Properly stock your project according to Competency #2, "Stocking a masonry project."
4. Observe all safety procedures.
5. Study the illustrations. Pay close attention to bond and coursing layout.
6. A small lead like this is often built in the center of a long wall to hold the line from sagging or being blown by strong winds.

Spreading bed joint on outside face shell.

Spreading bed joint on nearside face shell.

Buttering the head joint.

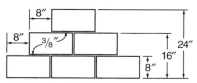

Front view rack back lead

Front view: Three-course rack-back lead.

12 Student Competency #4

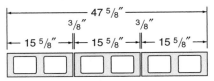

First course layout, Top view

Second course layout, Top view

Second and third course layout of rack-back lead.

PROCEDURE

1. Items needed:
   mortarboard/pan
   6 8″ × 8″ × 16″ hollow core block
   masonry tools
2. Draw the wall line on the floor 48″ long. Remember to keep 2′ between the near side of wall and your stocked material. Mark out block bond on this line with marks extending 1″ out toward you, so you can find them after spreading bed joint. Also extend each end of wall line 4″ so you can find that after spreading bed joint.
3. The first course of block shall be fully bedded, so spread the mortar on the floor just the same as spreading a brick bed joint. Spread the near side and furrow the bed joint, and then do the same for the far side.
4. Lay the block evenly and aligned with the wall line and bond marks. To apply the head joints, you may swipe them on the block in the wall or stand up the block you are going to lay and butter it on before picking it up to lay.
5. After laying three block in the first course, plumb each block and align with the wall line.
6. After plumbing the block, straightedge (range) the face of the block, remembering also to keep them even with the wall line.
7. Level the blocks with your four-foot level. Check to see if the top of the block is 8″ above the floor. Only check from one spot, because your floor may be slightly out of level. Tap lightly on the block with your trowel handle, or hammer where needed to make each sure each block is level and level with each other. After leveling, recheck plumb and alignment.
8. Spread the bed joint for the next course by swiping on the mortar. To do this, cut a trowel full of mud from the board and give it a quick, abrupt shake before you leave the board. This will spread the mortar uniformly on the trowel and help it stick to the trowel until you get the mud off. If you are right-handed, reach to the left end of the project and, keeping your trowel hand over the middle of the block, slide the forward tip of the trowel down and over the outside face shell of the block, pulling the trowel towards you. If you are left–handed, start on the outside right-hand end of the project. Continue until you have spread mortar on the outside face shell. Now use the same motion to spread the face shell towards you, but rotate the trowel handle in your hand so the trowel can swipe down and over the face shell toward you. Lay the two blocks in the second course and plumb, straight edge, and level them. Remember to check again for plumb and straight edge after you have leveled block. Check to see if the top of your block is 8″ above the top of the first course.
9. Lay the last block or third course just as you did in the previous courses. Don't forget to maintain plumb bond.
10. Tool your mortar joints when they are thumbprint hard and clean your area. When the project is done, clean your area and ask the instructor for an evaluation.

Instructor's Evaluation and Checklist

Student's Name: _____

	Possible Points	Points Off
1. Properly set up and stocked	10	
2. Safety procedures followed	10	
3. Plumb	10	
4. Level	10	
5. Height	10	
6. Bond	10	

7. Tooling:
   Holes
   Smears
   Tags or fingernails protruding from wall
   Uniform shape
	10	

8. Proper use of trowel and tools	10	
9. Motion study as described in procedure and good masonry practices	10	
10. Neatness	10	
	100	

TOTAL

FINAL GRADE

_____ SCORE

(Required to pass: Beginning 60–80%, Intermediate 80–90%, Advanced 90%)

Instructor's Comments:

Student Competency #5

TASK

Cutting masonry units with the brick hammer and brick set chisel.

Cutting brick with the brick hammer. Note that the mason's fingers and thumb are not placed on the side of the brick being scored.

PERFORMANCE OBJECTIVE

- The student shall be able to properly and safely cut a brick and a concrete block to the desired dimension with a brick hammer and the hammer and brick set chisel.

RELATED LEARNING ACTIVITIES

1. Read and review Unit 3 of the *Masonry Skills* textbook.
2. Watch the demonstration by the instructor.
3. Use safety glasses!
4. Skilled masons can quickly and efficiently cut masonry units by hand so production is not slowed down.

PROCEDURE

1. Items needed:
   2 common bricks
   2 two-core concrete blocks
   masonry tools
2. Mark the brick and block to the dimensions on project plan.
3. Cut the brick in half with the blade end of the brick hammer. *Remember:* Keep your wrist straight and bend only your elbow when cutting with the brick hammer.

16   Student Competency #5

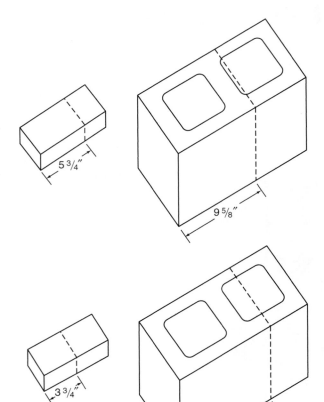

Cutting brick with the brick set. The flat side of the chisel must face the finished cut being made. Keep fingers above the cutting edge of the chisel to avoid injuries.

Cut lines.

4. Cut the block at the 11 5/8" mark. Remember to score the block and brick lightly all around before striking harder with your final cut.
5. Using the brick set chisel, cut the brick at the 5 3/4" mark.
6. Using the brick set chisel, cut the block at the 9 5/8" mark. Remember to score the block lightly before striking harder with the final cut.
7. Chippage of masonry units: Chips on the face that are 3/16" total or smaller shall be accepted when cutting concrete block and common brick by hand.
8. When the project is done, clean your area and ask the instructor for an evaluation.

Instructor's Evaluation and Checklist

Student's Name: _____

	Possible Points	Points Off
1. Proper safety practices and eye protection	20	
2. Half brick. Face is not chipped and within 1/4″ of dimension line on face.	20	
3. 11 5/8″ block. Face is not chipped and within 1/4″ of dimension line on face.	20	
4. 5 3/4″ brick. Face is not chipped and within 1/8″ of dimension line on face.	20	
5. 9 5/8″ block. Face is not chipped and within 1/8″ dimension line on face.	20	
	100	

TOTAL

FINAL GRADE

_____ SCORE

(Required to pass: Beginning 60–80%, Intermediate 80–90%, Advanced 90%)

Instructor's Comments:

SECTION TWO
DEVELOPMENTS AND MANUFACTURE OF BRICK AND CONCRETE MASONRY UNITS

SECTION TWO

CENTER OPERATIONS AND
MAINTENANCE OF FACILITIES,
GROUNDS, AND EQUIPMENT

Student Competency #6

TASK

Identifying and explaining the common ways to use ten masonry units.

PERFORMANCE OBJECTIVE

- The student will identify and explain the common use of five types of concrete masonry units and five types of clay brick.

RELATED LEARNING ACTIVITIES

1. Read and review Units 5–8 in the *Masonry Skills* textbook.
2. The mason must be able to lay masonry units of different shapes and textures.

PROCEDURE

1. Examine the masonry units provided by your instructor, which are marked with the numbers 1–10.
2. On the grade sheet in the workbook, write the name of each masonry unit and the typical use for each.

Student Competency #6

Instructor's Evaluation and Checklist

Student's Name: _____

Instructor shall provide various clay brick and concrete masonry units based on what is commonly used in area.

Type of Unit	Common Use	Possible Points	Points Off
1		10	
2		10	
3		10	
4		10	
5		10	
6		10	
7		10	
8		10	
9		10	
10		10	
		100	

TOTAL
FINAL GRADE

_____ SCORE
(Required to pass: Beginning 60–80%, Intermediate 80–90%, Advanced 90%)

Instructor's Comments:

SECTION THREE
LAYING BRICK TO A LINE AND BUILDING A BRICK CORNER

Student Competency #7

TASK

Mixing mortar.

PERFORMANCE OBJECTIVES

- The student shall identify the proper tools and equipment used to mix mortar by hand and power mixer.
- The student shall describe the proper mix proportion for practice mortar.
- The student shall properly and safely mix mortar by hand and with a power mixer.

RELATED LEARNING ACTIVITIES

1. Read and review Unit 9 of the *Masonry Skills* textbook, "Mixing Mortar."
2. Receive instruction and demonstration of gasoline- or electric-powered mortar mixer.
3. Know the proper mix of practice mortar and write it down in your notebook. The mix proportion is 1 part masonry lime or Carotex, 3 parts mason sand, and enough water to mix to a workable consistency. A typical batch of practice mortar will be $1/2$ bag masonry lime, 3 five-gallon pails of sand, and about 1 5-gallon pail of water.* (*Dampness and gradation of sand can vary greatly, so water has to be adjusted accordingly.)
4. *Safety reminder:* Always wear your safety glasses, hearing protection, and dust mask when using the power mixer. When mixing by hand, always wear your safety glasses.

Elevate the mortar box to mix by hand.

26 Student Competency #7

Adding the sand to the mixer. Sand is thrown on the grate, never into the mouth of the mixer. The worker is wearing safety goggles for mixing.

Dumping the mortar from the mixer into the wheelbarrow.

PROCEDURE

1. Mixing by hand with a mason's hoe and mortar box: Elevate mortar box with block at each corner so it is 8" to 16" off the floor. Shovel in half of the sand needed. Place the proper amount of masonry lime or Carotex on top of the sand that is already in the tub. Place the rest of the sand needed for the batch in the tub.
2. Standing at one end of the mortar box, chop through the dry ingredients and pull them to the end of the box towards you. Continue chopping through the pile and pulling toward you but not spilling the mix on the floor. Go to the other end of the mortar box and chop through the mix and pull towards you. This chopping through and pulling action will mix the dry ingredients.
3. When materials are blended from this mixing, add water to the mix and chop through and pull to mix in the water. Again, this chopping and pulling motion will mix the materials as you work one side and then the other. Mix and add water until you have the proper consistency to lay brick or block. Finish the mixing job by actually mixing the hoe through the mortar box for about 15 seconds. This will help it to be light and workable. Pull the finished mix to one end so you can shovel it out.
4. Mixing with the power mixer: *Caution: Never place your hand in a power mixer while it is running. Put out of gear, or if electric-powered, unplug if it is still plugged in.* All mixers should be shut off or unplugged before reaching in to clean it or to retrieve a dropped tool or bag.
5. Check the oil and gas levels in the mixer. If electric, make sure the cord is in proper condition and you have a properly grounded plug. Make sure the grate is working and is shut when mixing and adding ingredients.
6. Start the mixer and add about three-quarters of the water needed. Shovel in half the sand required. (You may also use buckets to measure, and these can be dumped through the grate.)
7. Carefully add the rest of the sand and water until you have achieved the proper consistency of mortar. (Listen to the sound of the motor or engine to hear if the mixer is laboring too hard; you may need to add a little more water. It is important not to jam the mixer by not having enough water.)
8. With the mixer still running, dump the mixed batch into a wheelbarrow or tub. Use a trowel to scrape the rim of the mixer off into the wheelbarrow or tub.
9. After the mixer is empty, put a small amount of water in and let the paddles rotate for a few seconds before shutting off. This will keep it clean for the next mix.
10. When the project is done, clean your area and ask the instructor for an evaluation.

Instructor's Evaluation and Checklist

Student's Name: _____

	Possible Points	Points Off
1. Mixing by hand:		
Mud box elevated	5	
Proper ingredients	5	
Proper mixing techniques with hoe	10	
Mortar is proper consistency.	10	
Safety glasses worn	10	
Tools are properly cleaned.	10	
2. Mixing with power mixer:		
Safety grate down	10	
Proper sequence and proportions of materials	10	
Mortar is proper consistency.	10	
Safety glasses, hearing protection, and ear protection worn	10	
Tools and mixer are properly cleaned.	10	
	100	

TOTAL

FINAL GRADE

_____ SCORE

(Required to pass: Beginning 60–80%, Intermediate 80–90%, Advanced 90%)

Instructor's Comments:

Student Competency #8

TASK

Laying brick to the line with corner poles or speed leads.

Position of the fingers when laying brick to the line. Notice that the mason's fingers grasp only the top edge of the brick.

PERFORMANCE OBJECTIVES

- The student shall properly mark out bond, coursing, and stock materials to lay the wall.
- The student shall be able to properly use the mason's line and line blocks with the corner poles that have been provided.
- The student shall know how to check the corner poles for plumb and face of wall setout and make the proper adjustments if needed.
- The student shall be able to maintain plumb jambs and proper plumb bond of head joints.

RELATED LEARNING ACTIVITIES

1. Properly stock your project according to Competency #2 of the workbook.
2. Observe all safety procedures.
3. Read and review Unit 10 of the *Masonry Skills* textbook.
4. Review previous projects in workbook.
5. Laying to the line is how most block and brick are installed on masonry jobs.

Student Competency #8

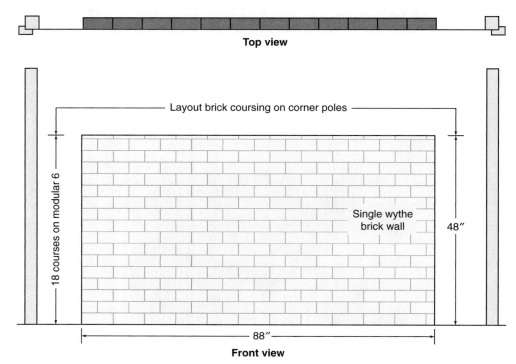

Front view and first course layout.

The correct spacing of brick is 1/16" from the line.

Adjusting brick to the line by pressing down with the hand.

PROCEDURE

1. Stock project area. Calculate how many brick are needed based on the corner pole setup suggested in the workbook or provided by your instructor.
2. Lay out bond on floor or footing.
3. Mark out brick coursing on the corner poles to the proper height.
4. Hang line on corner poles at the mark for the top of first course.
5. Using the procedures you learned in Competency #4 of the workbook, lay the first course of brick to the line.
6. Cut bricks in half with your hammer and use bats for half brick at each end of wall as needed.
7. Lay wall to proper height.
8. Tool mortar joints when they are thumbprint hard.
9. When the project is done, clean your area and ask the instructor for an evaluation.

Laying the closure brick in the wall. Notice that the mason has double jointed both the closure brick and the two bricks surrounding it. This action protects the wall against leaks.

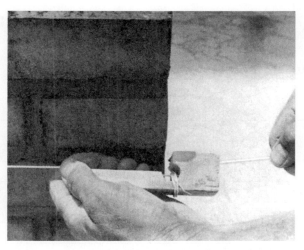

Wrapping line around line block and pulling it tight to place on brick corner.

Student Competency #8

Instructor's Evaluation and Checklist

Student's Name: _____

	Possible Points	Points Off
1. Project properly set up and stocked	10	
2. Safety procedures followed	10	
3. Plumb. Check five places.	10	
4. Level. Check top of wall.	10	
5. Height. Check proper coursing layout and uniform joints.	10	
6. Bond. Check plumb bond in two places.	10	
7. Tooling: 　Holes 　Smears 　Tags or fingernails protruding from wall or joint. 　Uniform shape	10	
8. Proper use of trowel and tools	10	
9. Motion study as described in procedure and good masonry practices	10	
10. Neatness	10	
	100	

TOTAL

FINAL GRADE

_____ SCORE
(Required to pass: Beginning 60–80%, Intermediate 80–90%, Advanced 90%)

Instructor's Comments:

Student Competency #9

TASK

Laying brick to the line and building a masonry opening using corner poles or speed leads.

PERFORMANCE OBJECTIVE

- The student shall be able to lay brick to the line using corner poles and build in a specified masonry opening with a flush rowlock sill in the wall.

RELATED LEARNING ACTIVITIES

1. Properly stock your project according to Competency #2 of the workbook.
2. Observe all safety procedures.
3. Watch the demonstration by the instructor.
4. Read and review Unit 10 of the *Masonry Skills* textbook.
5. Review related projects in workbook.
6. Masons must keep jambs plumb so window or door units fit.

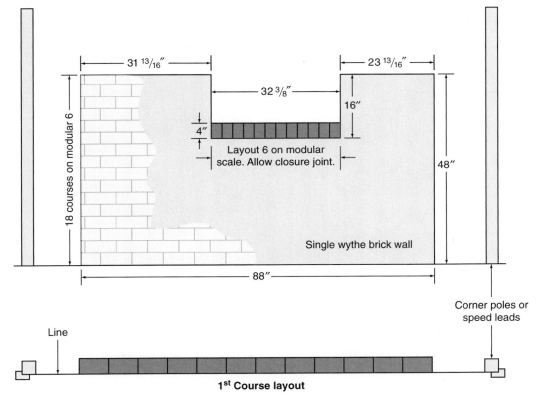

Front view and first course layout.

PROCEDURE

1. Stock project area. Calculate how many bricks are needed based on the corner pole project in this workbook unit or the corner pole setup provided by your instructor.
2. Lay out bond on the floor or footing.
3. Mark out brick coursing on corner poles.

4. Hang the line and lay brick according to procedure in Competencies #4 and #6 of this workbook.
5. Tool the joints as needed when they are thumbprint hard.
6. When you reach the course where the masonry opening starts, mark out the width of the opening at the proper location.
7. Continue to lay the wall with the masonry opening to the proper finished height. Be careful to plumb the jambs of the masonry opening and the end jambs.
8. After wall is finished and tooled, lay out brick spacing marks for the rowlock sill. Use the #6 scale on the mason's modular ruler. Make the marks lightly on the face of the brick just below the masonry opening. Allow for a closure joint at each end of sill layout.
9. Lay a rowlock brick at each end of the opening, keeping it flush with the face of the wall. Use the hammer to cut brick in half to make the rowlock brick.
10. Move your line blocks and line down to the rowlock level. Carefully twig the line on rowlock brick laid at each end. Carefully set a brick on each twig to hold line in place.
11. Lay the rowlock course, being careful to follow the layout marks, and butter full joints on the brick before laying them. When you get to the closure brick, make sure you butter the brick being laid and the edges of the brick in the opening.
12. Tool all joints.
13. When the project is done, clean your area and ask the instructor for an evaluation.

Instructor's Evaluation and Checklist

Student's Name: _____

	Possible Points	Points Off
1. Properly set up and stocked	10	
2. Safety procedures followed	10	
3. Plumb. Check six places.	10	
4. Level. Check top of wall and rowlock sill.	10	
5. Height. Check proper course layout and uniform joints.	10	
6. Bond. Check plumb bond in two places.	10	
7. Tooling: 　Holes 　Smears 　Tags or fingernails protruding from wall or joint 　Uniform shape	10	
8. Proper use of trowel and tools	10	
9. Motion study as described in procedure and good masonry practices. Measurement of masonry opening: Check one place.	10	
10. Neatness	10	
	100	

TOTAL

FINAL GRADE

_____ SCORE

(Required to pass: Beginning 60–80%, Intermediate 80–90%, Advanced 90%)

Instructor's Comments:

Student Competency #10

TASK

Laying brick veneer on running bond and a soldier course to the line with corner poles or speed leads.

PERFORMANCE OBJECTIVES

- The student shall properly mark out bond, coursing, and stock materials to lay the wall.
- The student shall be able to properly use the mason's line and line blocks with the corner poles that have been provided.
- The student shall know how to check the corner poles for plumb and face of wall setout, and make the proper adjustments if needed.
- The student shall be able to maintain plumb jambs and proper plumb bond of head joints.

RELATED LEARNING ACTIVITIES

1. Properly stock your project according to Competency #2, "Stocking a masonry project."
2. Observe all safety procedures.
3. Read and review Unit 10 of the *Masonry Skills* textbook.
4. Review previous projects in workbook.
5. Soldier brick add a pleasing design to the wall. Keep the soldier brick plumb.

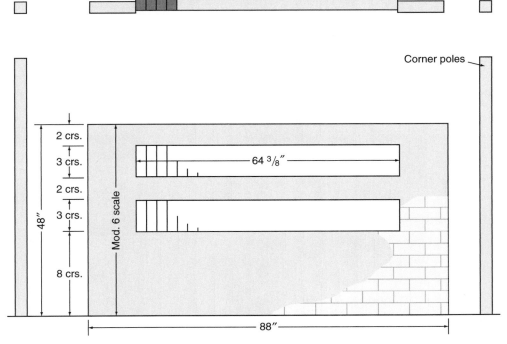

Front view.

PROCEDURE

1. Stock project area. Calculate how many bricks are needed_____ based on the project in the workbook or provided by your instructor.
2. Lay out dimension and bond on floor or footing.
3. Mark out brick coursing on the corner poles to the proper height.

4. Hang line on corner poles at the mark for the top of first course.
5. Using the procedures you learned in Competency #4 of the workbook, lay the first course of brick to the line.
6. Cut bricks in half with your hammer and use bats for half brick at each end of wall as needed. Keep brick face exposed on jambs. The end of the brick (bat) shall be the half brick in the face of the wall.
7. Lay eight courses. Remember to keep jambs plumb.
8. Raise the line for the ninth course. Then mark out the position of the soldier course. Use the modular ruler to lay out the bond for the soldier brick. Don't forget to allow a closure joint.
9. Lay the ninth, tenth, and eleventh courses. Keep jambs plumb. After completing the eleventh course, lay the soldier brick.
10. Lay the twelfth and eleventh courses. Raise the line and mark out the position and bond for the inset soldier course.
11. Lay the fourteenth, fifteenth, and sixteenth courses. Twig the line in from the face $1/2''$. Lay the soldier course to the line. Remember to keep the soldier brick plumb. The brick should be laid to the bottom of the mason's line, since it is twigged on top of the brick and not hung on the face.
12. Lay the last two courses.
13. Tool mortar joints when they are thumbprint hard. Use the flat slicker to tool the joints around the inset soldier brick. Use the concave jointer to tool between the soldier brick.
14. When the project is done, clean your area and ask the instructor for an evaluation.

Instructor's Evaluation and Checklist

Student's Name: _____

	Possible Points	Points Off	
1. Project properly set up and stocked	10		
2. Safety procedures followed	10		
3. Plumb. Check five places.	10		
4. Level. Check top of wall.	10		
5. Height: Check proper coursing layout and uniform joints.	10		
6. Bond. Check soldier course layout.	10		
7. Tooling. 　Holes 　Smears 　Tags or fingernails protruding from wall or joint 　Uniform shape	10		
8. Proper use of trowel and tools	10		
9. Motion study as described in procedure and good masonry practices	10		
10. Neatness	10		
	100		**TOTAL**
			FINAL GRADE

_____ **SCORE**

(Required to pass: Beginning 60–80%, Intermediate 80–90%, Advanced 90%)

Instructor's Comments:

Student Competency #11

TASK

Building a brick rack-back lead.

Laying a course in a rack-back lead. Note how the end bricks are racked back a half lap.

PERFORMANCE OBJECTIVE

■ The student shall properly lay out and build a 4" brick rack-back lead.

RELATED LEARNING ACTIVITIES

1. Properly stock your project according to Competency #2, "Stocking a masonry project."
2. Observe all safety practices.
3. Read and review Unit 11 of the *Masonry Skills* textbook.
4. Watch the demonstration by the instructor.
5. A lead like this may be built in the middle of a long wall to hold the line with a twig and to keep strong winds from blowing the line out of position.

PROCEDURE

1. Items needed:
   brick_____
   supply of shop mortar
   masonry tools
2. Lay out the project on the floor according to project plans. *Remember:* Extend the wall layout lines and bond lines so you can find them after you spread the bed joint. *Option:* Lay the project on a 2" × 4" × 48" board that is set up on two 4" block, using the front edge of the board as the face of the wall.
3. Spread the mortar for the first course and lay the brick on the bond marks. Level the brick with the four-foot level. Check each brick for plumb. Straightedge the brick to make sure they are all in. Recheck for level and proper height. Use the modular ruler to keep the courses even on the #6 scale.
4. Spread mortar and lay the second course. Be sure to start in 4" from the first course and keep on bond. Level plumb and straight edge using the first course as a guide.
5. Continue laying the next four courses, plumbing, leveling, and checking for height. Check plumb bond. The head joints should be plumb with the ones below on every other course.
6. When the project is done, clean your area and ask the instructor for an evaluation.

42 Student Competency #11

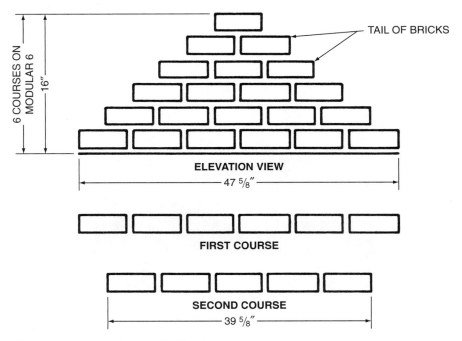

Constructing a 4" rack–back lead in the running bond.

Tooling the mortar joints with a convex sled runner.

Brushing the lead.

Tailing a rack–back lead. Notice that the plumb rule is in line with the corner of each top brick to assure proper alignment.

Completed rack–back lead built on a wooden 2" × 4".

Instructor's Evaluation and Checklist

Student's Name: _____

	Possible Points	Points Off	
1. Project set up and properly stocked	5		
2. Safety procedures followed	5		
3. Plumb. Check five places.	15		
4. Level. Check six places.	15		
5. Height. Check six places.	10		
6. Bond. Check plumb bond in three places.	10		
7. Tooling: 　Holes 　Smears 　Tags or fingernails that protrude from wall 　Uniform shape	10		
8. Proper use of trowel and tools	10		
9. Motion study as described in procedure and good masonry practices	10		
10. Neatness	10		
	100		**TOTAL**
			FINAL GRADE

_____ **SCORE**

(Required to pass: Beginning 60–80%, Intermediate 80–90%, Advanced 90%)

Instructor's Comments:

Student Competency #12

TASK

Building a brick corner.

Laying the first course of bricks in mortar. The mason is leveling the corner brick and the end brick before moving bricks in the center. This is the procedure to follow when working on a base that is not level.

PERFORMANCE OBJECTIVE

■ The student shall be able to lay out and build a brick corner nine courses high in the running bond.

RELATED LEARNING ACTIVITIES

1. Properly stock your project according to Competency #2, "Stocking a masonry project."
2. Observe all safety practices.
3. Read and review Unit 11 of the *Masonry Skills* textbook.
4. Watch the demonstration by the instructor.
5. A mason builds the brick corner as a guide for the masons to build the rest of the wall.

46 Student Competency #12

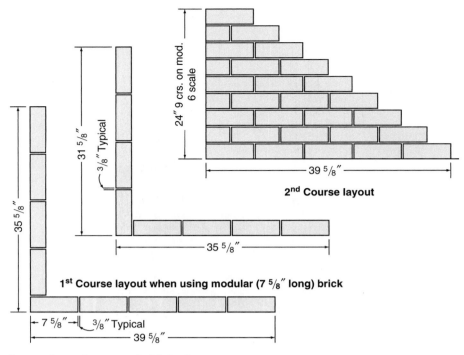

Layout for a nine course brick lead.

Dry bonding the proper number of bricks for a corner nine courses high. Notice that there are five bricks leading in one direction and four in the other direction.

Reversing the procedure and leveling the opposite side. The mason has not yet disturbed any bricks between the leveling points.

PROCEDURE

1. Items needed:
   brick____
   supply of shop mortar
   masonry tools
2. Lay out the project lines and bond marks on the floor or foundation. *Remember:* Extend the wall layout lines and the bond lines so you can see them after you spread the bed joint. Use the modular scale to build the brick courses to the #6 scale.
3. Spread the mortar for one side of the corner. Lay that course of bricks. Plumb, straightedge and level. Be careful to stay on bond. Spread mortar for the other side of the corner and lay that course of brick. Plumb,

Checking the corner on a diagonal to be sure that the first course is level.

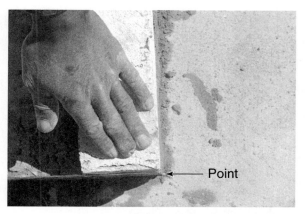

Sighting down the outermost point of the corner bricks to be certain they are plumb. The corner shown, five courses high, is in perfect alignment.

Striking the corner with a V-jointer. Mortar joints should be tooled before they are too stiff.

straight edge, and level. *Remember:* Always go back and check your layout, plumb, level, straight edge, bond, and height before moving on to the next course.

4. Lay the second course in the same manner. Spread the mortar for one side. Lay that side of the corner and check for plumb, level, straight edge and height. When you lay the other side of the corner, be careful not to dislodge the brick already laid. Use the first course as your guide to plumb the corner and keep bond.
5. Continue to lay each course as described. After three courses, you will be able to spread the full bed joint on the corner and have time to lay and adjust the brick on that course before the mortar sets too much.
6. Tool the joints when thumbprint hard.
7. When the project is done, clean your area and ask the instructor for an evaluation.

Instructor's Evaluation and Checklist

Student's Name: _____

	Possible Points	Points Off
1. Project set up and properly stocked	5	
2. Safety procedures followed	5	
3. Plumb. Check six places.	15	
4. Level. Check nine places.	15	
5. Height. Check nine places.	10	
6. Bond. Check plumb bond in four places.	10	
7. Tooling: Tool face side only. Cut mud from the back. Holes Smears Tags or fingernails that protrude from wall Uniform shape	10	
8. Proper use of trowel and tools	10	
9. Motion study as described in procedure and good masonry practices	10	
10. Neatness	10	
	100	

TOTAL

FINAL GRADE

_____ SCORE

(Required to pass: Beginning 60–80%, Intermediate 80–90%, Advanced 90%)

Instructor's Comments:

Student Competency #13

TASK

Laying brick corners and filling in the wall.

PERFORMANCE OBJECTIVE

■ The student shall be able to properly lay out and build two brick corners, and lay the wall in with the line.

RELATED LEARNING ACTIVITIES

1. Properly stock your project according to Competency #2, "Stocking a masonry project."
2. Observe all safety practices.
3. Read and review Unit 11 of the *Masonry Skills* textbook.
4. Watch the demonstration by the instructor.
5. Review Competency #6 of this workbook, "Laying brick to the line."

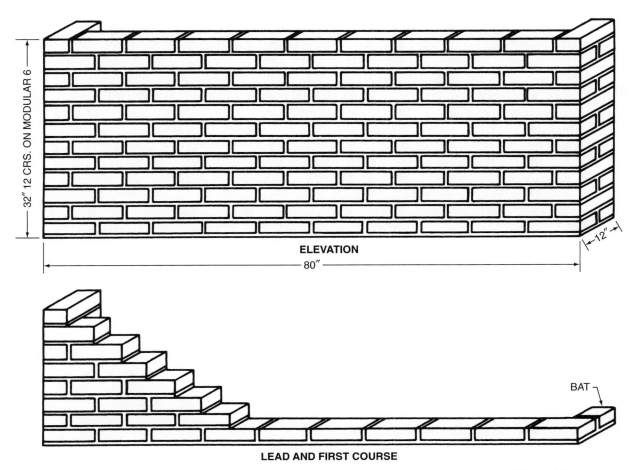

Laying a brick corner and building a wall in the running bond with a line. Dimensions are for modular brick.

PROCEDURE

1. Items needed:
   brick _____
   supply of shop mortar
   masonry tools
2. Lay out the project on the floor or foundation. Mark out the bond. To maintain the proper height, you will use the modular ruler to lay the brick courses on the #6 scale.
3. Lay four brick in the first course at each end of the project. Check for proper height and make sure they are level with each other. If the floor or foundation is out of level, it is important to get the first course of brick level. Use the transit or a long straightedge to check for level.
4. Use the mason's line and line blocks to check the "range" of each corner to make sure they line up with each other. Be careful not to tighten the line too much and pull the brick out of position.
5. Lay the corner lead of one side of the project six courses high. Use a bat for the half brick the corner jamb requires.
6. Lay the other side of the project six courses high. Check the height from the top of the first course.
7. When the six course leads are finished, hang the mason's line and lay in the brick wall.
8. Build the corner leads six more courses and finish the wall.
9. Tool joints when thumbprint hard.
10. When the project is done, clean your area and ask the instructor for an evaluation.

Instructor's Evaluation and Checklist

Student's Name: _____

	Possible Points	Points Off
1. Project set up and properly stocked	5	
2. Safety procedures followed	5	
3. Plumb. Check eight places.	15	
4. Level. Check three places.	10	
5. Height. Check twelve courses at two positions.	15	
6. Bond: Check running bond. Check plumb bond in three places.	10	
7. Tooling: Tool face side only. Cut mud from back. Holes Smears Tags or fingernails that protrude from wall Uniform shape	10	
8. Proper use of trowel and tools	10	
9. Motion study as described in project and good masonry practices	10	
10. Neatness	10	
	100	

TOTAL

FINAL GRADE

_____ **SCORE**

(Required to pass: Beginning 60–80%, Intermediate 80–90%, Advanced 90%)

Instructor's Comments:

SECTION FOUR
MORTAR AND ESSENTIALS OF BONDING

Student Competency #14

TASK

Laying brick to the line with quoin corners using corner poles or speed leads.

PERFORMANCE OBJECTIVE

- The student shall be able to lay brick to the line using corner poles and build in quoin corners at each end of the wall.

RELATED LEARNING ACTIVITIES

1. Properly stock your project according to Competency #2, "Stocking a masonry project."
2. Observe all safety procedures.
3. Watch the demonstration by the instructor.
4. Read and review Unit 10 of the *Masonry Skills* textbook.
5. Review related projects in the workbook.
6. Quoin corners take a little more time, but they add elegance to a flat wall or corner.

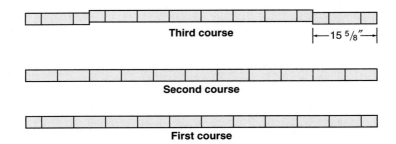

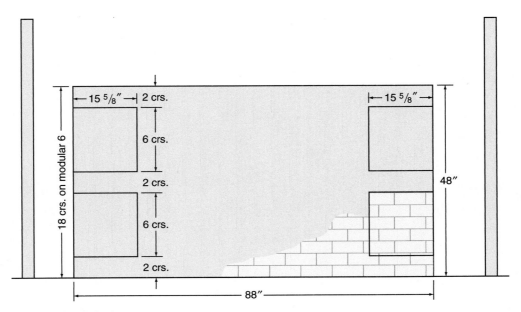

Front view and beginning course layout.

PROCEDURE

1. Stock project area. Calculate how many bricks and/or block are needed for your project based on the corner pole project in this workbook or the corner pole setup provided by your instructor.
2. Lay out bond on the floor or footing.
3. Mark out the brick coursing on the corner pole.
4. Hang the line and lay the brick according to procedures in Competencies #4 and #6 of this workbook.
5. Tool the joints as needed when thumbprint hard.
6. Lay the first two stretcher courses through. When you lay the third course, do not lay the first two bricks at each end of the wall. Lay the stretcher brick in between to the line. Be sure to stay on bond and allow a 16″ space for the quoin brick.
7. After laying the third course to the line, move the line up to the next course and lay the two quoin brick by corbelling out $3/4″$. Plumb, level, and range quoin brick by hand.
8. Continue to lay the next five courses this way, going back each time to lay the quoin brick in by hand after you have laid the stretcher brick to the line.
9. After laying the six-course quoin corner, lay two stretcher courses.
10. Repeat building the next six courses with quoin corners.
11. Finish the project with the final two stretcher courses.
12. Tool all joints when thumbprint hard. Use the flat slicker under, beside, and on top of the quoin corners.
13. When the project is done, clean your area and ask the instructor for an evaluation.

Instructor's Evaluation and Checklist

Student's Name: _____

	Possible Points	Points Off
1. Properly set up and stocked	5	
2. Safety procedures followed	5	
3. Plumb. Check seven places.	15	
4. Level. Check top of wall and top and bottom of each quoin.	15	
5. Height. Check proper course layout and uniform joints.	10	
6. Bond. Check plumb bond in two places.	10	
7. Tooling: Holes Smears Tags or fingernails that protrude from wall Uniform shape	10	
8. Proper use of trowel and tools	10	
9. Motion study as described in procedure and good masonry practices	10	
10. Neatness	10	
	100	

TOTAL

FINAL GRADE

_____ SCORE

(Required to pass: Beginning 60–80%, Intermediate 80–90%, Advanced 90%)

Instructor's Comments:

Student Competency #15

TASK

Laying brick corners and filling in the wall with common bond.

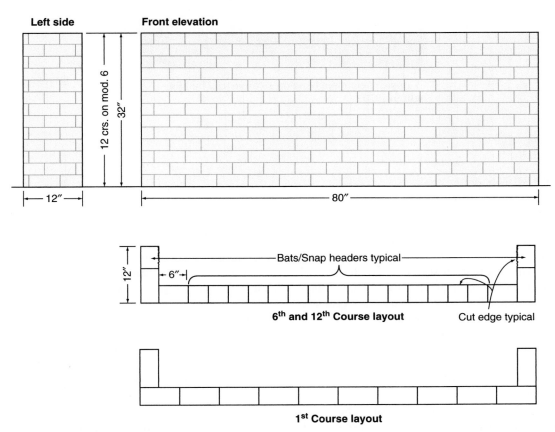

Brick veneer laid in common bond to duplicate a historic bond pattern. Use hammer cut bats for the header courses.

PERFORMANCE OBJECTIVES

- The student shall be able to properly lay out and build two brick corners, and lay the wall in with the line using common bond.
- The student shall make the proper cuts for the corner leads to achieve common bond.

RELATED LEARNING ACTIVITIES

1. Properly stock your project according to Competency #2, "Stocking a masonry project."
2. Observe all safety practices.
3. Read and review Units 11 and 15 of the *Masonry Skills* textbook.
4. Watch the demonstration by the instructor.
5. Review Competency #6 of this workbook, "Laying brick to the line."
6. This project is typical of historic bond patterns used in brick veneer.

Example of common bond on an industrial building built in 1920.

PROCEDURE

1. Items needed:
   brick _____
   supply of shop mortar
   masonry tools
2. Lay out the project on the floor or foundation. Mark out the bond of the first course, which shall be running bond. To maintain the proper height, you will use the modular ruler to lay the brick courses on the #6 scale.
3. Lay four brick in the first course at each end of the project. Check for proper height and make sure they are level with each other. If the floor or foundation is out of level, it is important to get the first course of brick level. Use the transit or a long straightedge to check for level.
4. Use the mason's line and line blocks to check the "range" of each corner to make sure they line up with each other. Be careful not to tighten the line too much and pull the brick out of position.
5. Lay the corner lead of one side of the project five courses high. Use a bat for the half brick the corner jamb requires.
6. Lay the other side of the project five courses high. Check the height from the top of the first course at the corners only.
7. When the first five courses are finished, hang the mason's line and lay in the brick wall.
8. The sixth course shall be the header course. Cut two three-quarter brick (Dutch corner) and dry lay the sixth course as on the project plan. Make sure the bats or snap headers are centered over the head joints of the course below. When the layout is proper, lay a six-course lead at each end, and then hang the line and fill in. Remember the sixth course is the header course.
9. Build the corners six more courses and finish the wall. Remember the sixth and twelfth courses are the header courses.
10. Tool joints when thumbprint hard.
11. When the project is done, clean your area and ask the instructor for an evaluation.

Instructor's Evaluation and Checklist

Student's Name: _____

	Possible Points	Points Off
1. Project set up and properly stocked	5	
2. Safety procedures followed	5	
3. Plumb. Check eight places.	15	
4. Level. Check three places.	10	
5. Height. Check twelve courses at two positions.	15	
6. Bond. Check common bond. Check plumb bond in three places.	10	
7. Tooling: Tool face side only. Cut mud from back. Holes Smears Tags or fingernails that protrude from wall Uniform shape	10	
8. Proper use of trowel and tools. Cuts are accurate to $1/8''$ plus or minus.	10	
9. Motion study as described in project and good masonry practices	10	
10. Neatness	10	
	100	

TOTAL

FINAL GRADE

_____ SCORE

(Required to pass: Beginning 60–80%, Intermediate 80–90%, Advanced 90%)

Instructor's Comments:

Student Competency #16

TASK

Laying brick corners and filling in the wall with English bond.

Brick wall laid in common bond with a Flemish bond header every seventh course.

PERFORMANCE OBJECTIVES

- The student shall be able to properly lay out and build two brick corners, and lay the wall in with the line using English bond.
- The student shall make the proper cuts for the corner leads to achieve English bond.

RELATED LEARNING ACTIVITIES

1. Properly stock your project according to Competency #2, "Stocking a masonry project."
2. Observe all safety practices.
3. Read and review Units 11 and 15 of the *Masonry Skills* textbook.
4. Watch the demonstration by the instructor.
5. Review Competency #6 of this workbook, "Laying brick to the line."
6. This project is typical of historic bond patterns used in modern brick veneer.

Student Competency #16

Left side **Front elevation**

12 crs. on mod. 6
32"

|— 12" —| |————————— 80" —————————|

Snap headers   Cut edge
12" |← 6" →| |← 6" →|
2nd Course

1st Course

Brick veneer laid in English bond to duplicate an historic bond pattern. Great care must be used to maintain plumb bond of head joints.

PROCEDURE

1. Items needed:
   brick _____
   supply of shop mortar
   masonry tools
2. Lay out the project on the floor or foundation. Mark out the bond of the first course, which shall be running bond. To maintain the proper height, you will use the modular ruler to lay the brick courses on the #6 scale.
3. Lay four brick in the first course at each end of the project. Check for proper height and make sure they are level with each other. If the floor or foundation is out of level, it is important to get the first course of brick level. Use the transit or a long straightedge to check for level.
4. Use the mason's line and line blocks to check the "range" of each corner to make sure they line up with each other. Be careful not to tighten the line too much and pull the brick out of position.
5. Lay the first course of brick for the project in the running bond.
6. Cut two three-quarter brick (Dutch corners) for each corner and dry lay the second course as shown on the project plan.
7. When you are sure of the proper layout, make enough three-quarter cuts to build the leads five courses high. Lay the corner lead of one side of the project five courses high. Use a bat for the half brick the corner jamb requires.
8. Lay the other lead five courses high. Check the height from the top of the first course at the corners only.
9. When the two corner leads are finished, hang the mason's line and lay in the brick wall. Remember every other course is a header course.
10. When laying in the wall, make sure the bats or snap headers are centered over the head joints of the course below.
11. Build the corners six more courses and finish the wall. Remember every other course is a header course.
12. Tool joints when thumbprint hard.
13. When the project is done, clean your area and ask the instructor for an evaluation.

Student Competency #16

Instructor's Evaluation and Checklist

Student's Name: _____

	Possible Points	Points Off
1. Project set up and properly stocked	5	
2. Safety procedures followed	5	
3. Plumb. Check eight places.	15	
4. Level. Check three places.	10	
5. Height. Check courses at two positions.	15	
6. Bond. Check common bond. Check plumb bond in three places.	10	
7. Tooling: Tool face side only. Cut mud from back. Holes Smears Tags or fingernails that protrude from wall Uniform shape	10	
8. Proper use of trowel and tools. Cuts are accurate to $1/8''$ plus or minus.	10	
9. Motion study as described in project and good masonry practices	10	
10. Neatness	10	
	100	

TOTAL

FINAL GRADE

_____ SCORE

(Required to pass: Beginning 60–80%, Intermediate 80–90%, Advanced 90%)

Instructor's Comments:

Student Competency #17

TASK

Laying brick corners and filling in the wall with Flemish bond.

PERFORMANCE OBJECTIVES

- The student shall be able to properly lay out and build two brick corners, and lay the wall in with the line using Flemish bond.
- The student shall make the proper cuts for the corner leads to achieve Flemish bond.

RELATED LEARNING ACTIVITIES

1. Properly stock your project according to Competency #2, "Stocking a masonry project."
2. Observe all safety practices.
3. Read and review Units 11 and 15 of the *Masonry Skills* textbook.
4. Watch the demonstration by the instructor.
5. Review Competency #6 of this workbook, "Laying brick to the line."
6. This project is typical of bond patterns used in brick veneer.

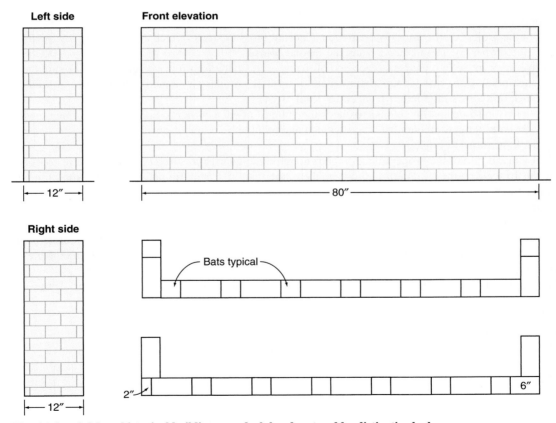

Flemish bond. Many historical buildings use dark headers to add a distinctive look.

Corner of a brick building built with Flemish bond. Note the 2″ starter pieces near the corner.

PROCEDURE

1. Items needed:
   brick _____
   supply of shop mortar
   masonry tools
2. Lay out the project on the floor or foundation. Cut four three-quarters (Dutch corners). Dry lay the first course. Verify the proper layout and then dry lay the second course.
3. Lay five brick in the first course at each end of the project. Check for proper height and make sure they are level with each other. If the floor or foundation is out of level, it is important to get the first course of brick level. Use the transit or a long straightedge to check for level.
4. Use the mason's line and line blocks to check the "range" of each corner to make sure they line up with each other. Be careful not to tighten the line too much and pull the brick out of position.
5. Lay the first course of brick.
6. When you are sure of the proper layout, make enough three-quarter cuts to build the leads five courses high. Use the #6 scale on the modular ruler to keep the proper height. Check for height from the top of the first course in the corner only.
7. Lay the corner lead of one side of the project five courses high. Use a bat for the half brick the corner jamb requires.
8. Lay the other lead five courses high. Check the height from the top of the first course at the corners only.
9. When the two corner leads are finished, hang the mason's line and lay in the brick wall. Remember every other brick is a header.
10. When laying in the wall, make sure the bats or snap headers are centered over the head joints of the course below.
11. Build the corners six more courses and finish the wall.
12. Tool joints when thumbprint hard.
13. When the project is done, clean your area and ask the instructor for an evaluation.

Instructor's Evaluation and Checklist

Student's Name: _____

	Possible Points	Points Off
1. Project set up and properly stocked	5	
2. Safety procedures followed	5	
3. Plumb. Check eight places.	15	
4. Level. Check three places.	10	
5. Height. Check twelve courses at two positions.	15	
6. Bond. Check common bond. Check plumb bond in three places.	10	

7. Tooling: Tool face side only. Cut mud from back.
    Holes
    Smears
    Tags or fingernails that protrude from wall
    Uniform shape
	10	

8. Proper use of trowel and tools. Cuts are accurate to $1/8''$ plus or minus.	10	
9. Motion study as described in project and good masonry practices	10	
10. Neatness	10	
	100	

TOTAL

FINAL GRADE

_____ SCORE

(Required to pass: Beginning 60–80%, Intermediate 80–90%, Advanced 90%)

Instructor's Comments:

Student Competency #18

TASK

Laying a 4″ block pier with corbelled brick top.

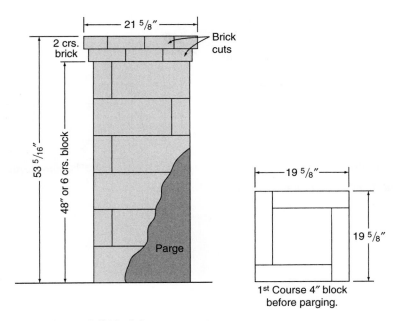

Front view and 4″ block layout.

PERFORMANCE OBJECTIVES

- The student shall properly lay out and build a 20″ × 20″ pier six courses high using 4″ block.
- The student shall lay two corbelled brick courses on top and parge the exterior of the 4″ block pier.
- The student shall be able to maintain plumb, level, range, and square on all four sides of the project.

RELATED LEARNING ACTIVITIES

1. Properly stock your project according to Competency #2, "Stocking a masonry project."
2. Observe all safety practices.
3. Review Unit 17 of the *Masonry Skills* textbook. The techniques for building a block pier are very similar to building a block corner because it is all level work.
4. Watch the demonstration by the instructor.

Student Competency #18

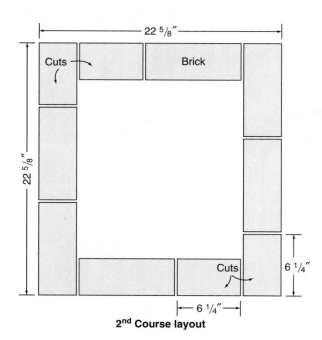

1st Course layout

2nd Course layout

Make neat accurate cuts to make corbelling look good.

PROCEDURE

1. Items needed:
   4″ block_____
   brick_____
   supply of shop mortar
   masonry tools
2. Lay out the project lines on the floor or foundation.
3. Lay the first course and double-check the plumb, level, square, and height. This will be your guide for the rest of the pier.
4. Lay the rest of the courses using caution, because 4″ block have a tendency to tip or lean. Do not tool the joints in the block work but cut them off inside and out. Be careful not to drop excess mortar in the cavity.
5. After six courses of block, start the brick by dry laying the first course according to the project plans. You will have to make some cuts. Follow the bond plan in the project plans. This design eliminates small pieces.
6. Lay first brick course with care so it doesn't tip. Holding your brick ruler sideways under the brick is a handy gauge to measure $3/4''$.
7. Lay the second course of brick and tool when thumbprint hard.
8. After the brick are laid, parge on a coat of mortar on the exterior of the 4″ block. Use a plastering trowel or your brick trowel. Cover the entire block and don't leave holes. You have the option of troweling a smooth finish or leaving it slightly rough.
9. After the project is done, clean your area and ask your instructor for an evaluation.

Instructor's Evaluation and Checklist

Student's Name: _____

	Possible Points	Points Off
1. Project set up and properly stocked	5	
2. Safety procedures followed	5	
3. Plumb. Check blocks in four places inside pier. Check brick in eight places.	15	
4. Level. Check top of brick and under first course.	10	
5. Height. Check top of pier and bricks laid on #6 scale modular ruler.	15	
6. Bond. Check for proper cuts in brickwork. Check for proper bond according to project plan. Check for uniform appearance when viewing from a distance of 5′.	10	
7. Tooling: Holes Smears Tags or fingernails that protrude from wall Uniform shape	10	
8. Proper use of trowel and tools. Parging covers all block and is finished neatly under brick.	10	
9. Motion study as described by procedure and good masonry practices	10	
10. Neatness	10	
	100	

TOTAL

FINAL GRADE

_____ SCORE

(Required to pass: Beginning 60–80%, Intermediate 80–90%, Advanced 90%)

Instructor's Comments:

Student Competency #19

TASK

Laying brick corbelling and coping.

PERFORMANCE OBJECTIVE

■ The student shall be able to properly lay out and build a composite wall with corbelled brick dentil work, sawtooth work, running bond, and 4″ block coping.

RELATED LEARNING ACTIVITIES

1. Properly stock your project according to Competency #2, "Stocking a masonry project."
2. Observe all safety practices.
3. Read and review Unit 26 of the *Masonry Skills* textbook.
4. Watch the demonstration by the instructor.
5. Block work has a cut. Maintain proper bond in block work. Avoid using cuts smaller than 8″.

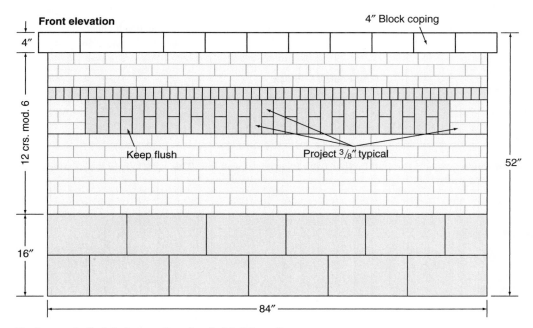

Various corbelled designs and coping finish this wall.

Next to last course of brick

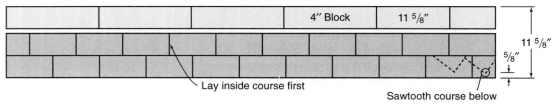

Lay inside course first
Sawtooth course below

Sawtooth course

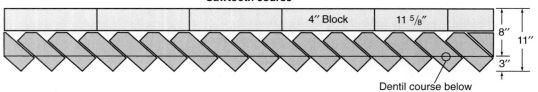

Dentil course below

Top of dentil course, 9 crs. brick

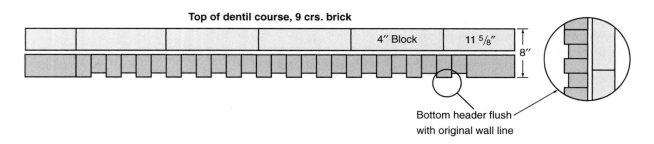

Bottom header flush with original wall line

First course 8″ block

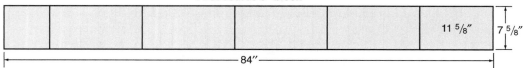

Each corbelled course projects the wall further from its original wall line. This style helped keep water from running down the face of the wall.

PROCEDURE

1. Items needed:
   8″ block_____
   8″ half block_____
   4″ block_____
   4″ half block
   wall ties or joint reinforcement
   supply of shop mortar
   masonry tools
2. Lay out project according to project plans.
3. Notice the placement of the cut block on second course. Lay the first two courses of block. Fill cores with masonry debris or cover with grout screen or hardware cloth.
4. Lay the first course with 4″ block and three courses of brick together so you can install wall ties. Build a lead at each end and lay in with line.
5. Build a six-course lead at each end of the project, including the $^3/_8″$ projected corners for the dentil work.
6. Hang the line to fill in the three courses of running bond.
7. Hang the line on top of the corner to lay the soldiers and headers in the dentil work. To help hold the corner leads, set a brick on top and back against the 4″ block to counterweight the corner.

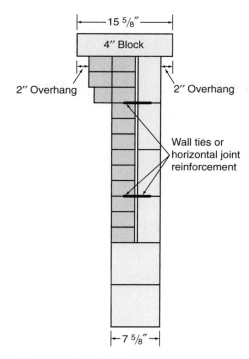

Side view of project 19.

8. Lay out the brick spacing for the dentil course in between the corner leads. Use the brick spacing ruler.
9. Lay the dentil course. Remember that the bottom header is flush with the original wall line. If the soldiers are tipping, carefully set a dry laid brick on top and against the 4″ block. Use a flat slicker to tool the inset dentils.
10. Dry lay the sawtooth course. Cut each end as required. You will have to clip or cut the back of the brick to fit. Once layout is adjusted and correct, lay the sawtooth course. Use a flat slicker to tool sawtooth brick.
11. Lay the last courses of brick, which are two wythes. Lay the inner wythes first to counterbalance the corbelled brick work.
12. Tool joints when thumbprint hard.
13. Lay out the coping (4″ × 8″ × 16″ block) using 2″ overhang on all sides and ends.
14. Lay a coping piece (block) at each end plumb and level both ways. Hang a line front and back and lay the coping pieces. Tool when thumbprint hard. Use flat slicker under the coping.
15. When you are done with your project, clean the area and ask your instructor for an evaluation.

Instructor's Evaluation and Checklist

Student's Name: _____

	Possible Points	Points Off
1. Project set up and properly stocked	5	
2. Safety procedures followed	5	
3. Plumb. Check four places.	10	
4. Level. Check seven places.	10	
5. Height. Check overall and coursing two places.	10	
6. Bond. Check block bond and corbelled brick for proper layout.	15	
7. Tooling: Holes Smears Tags or fingernails that protrude from wall Uniform shape	15	
8. Proper use of trowel and tools.	10	
9. Motion study as described by procedure and good masonry practices. Proper layout of brick design and corbel overhangs.	10	
10. Neatness.	10	
	100	

TOTAL

FINAL GRADE

_____ SCORE

(Required to pass: Beginning 60–80%, Intermediate 80–90%, Advanced 90%)

Instructor's Comments:

SECTION FIVE
LAYING CONCRETE BLOCK

Student Competency #20

TASK

Laying block to the line.

Laying the block gently in position so that it does not sink below the line. When properly positioned, it is even with top of the line and approximately $^1/_{16}''$ back from line.

PERFORMANCE OBJECTIVE

■ The student shall be able to set up and mark corner poles to lay an 8″ block wall that is 88″ long and 48″ tall.

RELATED LEARNING ACTIVITIES

1. Properly stock your project according to Competency #2, "Stocking a masonry project."
2. Observe all safety practices.
3. Read and review Unit 16 of the *Masonry Skills* textbook.
4. Watch the demonstration by the instructor.
5. Review block laying and mortar spreading practices in Competency #5 of this workbook.

PROCEDURE

1. Items needed:
   supply of shop mortar
   8″ block_____
   8″ jamb block_____
   8″ half block_____
   masonry tools
2. Mark the corner poles with the proper coursing. Be careful to check the floor or footing for level and make proper adjustments with the first course.
3. Mark out bond on the floor or footing.
4. Hang the mason's line on corner poles at the marks for the first course.
5. Lay the first course to the line. Remember to keep $^1/_{16}''$ away from line so you do not push the wall out of line (range).

Position of line and wall if you are using corner poles attached for 4″ veneer.

Brick veneer position

8″ Block

First course layout

Top view Work area

Mark corner poles or speed leads

6 crs. on 2 scale of modular ruler

48″

8″

The student may use corner poles set up for brick veneer by laying the block on the other side of line.

Block laid correctly to line, even with top of line and 1/16″ back from line.

Mortar head joints are applied to both ends of adjoining blocks to help form a more full joint for closure block.

6. After laying the first course, check each jamb end for plumb and alignment with layout marks.
7. Continue this procedure to lay the remaining five courses of block.
8. Be careful to plumb the jamb at each end of wall, keeping the bottom edge of block you are laying flush with block below and the top edge $1/16″$ away from line, and tool joints when thumbprint hard.
9. When the project is done, clean your area and ask the instructor for an evaluation.

Instructor's Evaluation and Checklist

Student's Name: _____

	Possible Points	Points Off
1. Project set up and properly stocked	5	
2. Safety procedures followed	5	
3. Plumb. Check five places.	15	
4. Level. Check top of each course.	15	
5. Height. Check proper course lay out and uniform bed joints.	10	
6. Bond. Check for running bond. Check plumb bond in one place.	10	
7. Tooling: Holes Smears Tags or fingernails protruding from wall Uniform shape	10	
8. Proper use of trowel and tools	10	
9. Motion study as described in procedure and good masonry practices	10	
10. Neatness	10	
	100	

TOTAL

FINAL GRADE

_____ SCORE

(Required to pass: Beginning 60–80%, Intermediate 80–90%, Advanced 90%)

Instructor's Comments:

Student Competency #21

TASK

Laying block to the line and building in a masonry opening with rowlock sill.

PERFORMANCE OBJECTIVES

- The student shall be able to set up and mark corner pole to build an 8″ block wall that is 88″ long and 48″ high.
- The student shall be able to build in a masonry opening with a rowlock sill.

RELATED LEARNING ACTIVITIES

1. Properly stock your project according to Competency #2, "Stocking a masonry project."
2. Observe all safety practices.
3. Read and review Unit 16 of the *Masonry Skills* textbook.
4. Watch the demonstration by the instructor.
5. Review block laying and mortar spreading practices in Competency #5 of this workbook.

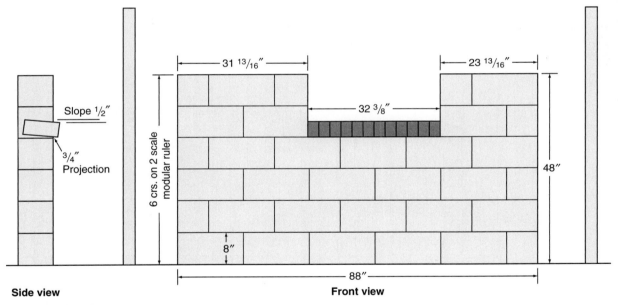

Side view

Front view

Front and side view.

PROCEDURE

1. Items needed:
   supply of shop mortar
   8″ block_____
   8″ jamb block_____
   8″ half block
   brick (in dry condition)_____
   masonry tools
2. Mark the corner poles with the proper coursing. Be careful to check the floor or footing for level and make proper adjustments with the first course of block.
3. Mark out bond on the floor or footing.

4. Hang the mason's line on the corner poles at the marks for the first course.
5. Lay the first course to the line. Remember to keep $1/16''$ away from the line to avoid pushing out of line (range).
6. After laying the first course, check each jamb end for plumb and alignment with the layout marks.
7. Continue this procedure until you have four courses laid.
8. Hang the line for the fifth course and mark out the masonry opening according to the project plan.
9. Lay the remaining two courses on each side of the masonry opening with the line. Remember to plumb all jambs, including the masonry opening.
10. When all block are laid, tool the joints and lay out the marks for the rowlock sill according to the project plans. Lightly mark the face of the block just below the rowlock sill location.
11. Fill the cells of the 8″ block in the masonry opening with masonry debris or cover with grout screen or hardware cloth.
12. Butter a mortar joint on the side of a brick and lay it in the rowlock position against the jamb of the masonry opening. To lay a sloped rowlock sill, butter the brick before setting in wall. Keep the brick level as you lay it into the bed joint. Just as you are releasing the brick from your hand, rotate it forward to give it the slope. Project the bottom of the brick $3/4''$ from the outside face of the wall. Lay another rowlock brick in the same manner at the other end of the opening.
13. Move the line blocks to the approximate height and location of the face of the rowlock sill. Carefully twig the line on the top outer edge of the two sill bricks laid. Be careful when setting a "holder" brick to hold twig in place. You do not want to dislodge sill bricks spotted at each end.
14. Lay the sill working from each end to the middle. Lay the front of the sill to the line. Level the back of the sill every third brick.
15. For the closure brick, butter head joints on each side of the brick you are laying and butter head joints on the adjoining brick.
16. Tool all joints when thumbprint hard. Use a flat slicker under the sill.
17. When the project is done, clean your area and ask the instructor for an evaluation.

Instructor's Evaluation and Checklist

Student's Name: _____

	Possible Points	Points Off
1. Project set up and properly stocked	5	
2. Safety procedures followed	5	
3. Plumb. Check six places.	15	
4. Level. Check top of each course and the front and back of sill.	10	
5. Height. Check proper course layout, uniform bed joints, and the proper slope on the sill.	15	
6. Bond. Check for running bond. Check for proper brick spacing in rowlock sill.	10	
7. Tooling: Holes Smears Tags or fingernails that protrude out from wall Uniform shape	10	
8. Proper use of trowel and tools	10	
9. Motion study as described in procedure and good masonry practices	10	
10. Neatness	10	
	100	

TOTAL

FINAL GRADE

_____ SCORE

(Required to pass: Beginning 60–80%, Intermediate 80–90%, Advanced 90%)

Instructor's Comments:

Student Competency #22

TASK

Laying 4″ × 8″ × 16″ split face block veneer to the line.

PERFORMANCE OBJECTIVES

- The student shall be able to set up and mark corner pole and lay 4″ split face block as a veneer on a wall that is 80″ long and 48″ high.
- The student shall properly spread a full bed joint and properly apply full head joints without dropping excess mortar.

RELATED LEARNING ACTIVITIES

1. Properly stock your project according to Competency #2, "Stocking a masonry project."
2. Observe all safety practices.
3. Read and review Unit 16 of the *Masonry Skills* textbook.
4. Watch the demonstration by the instructor.
5. Review block laying and mortar spreading practices in Competencies #4 and #5 of this workbook.
6. Split face block add an interesting textured finish to walls. *Be careful not to smear the face; it is hard to clean.*

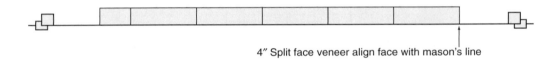

4″ Split face veneer align face with mason's line

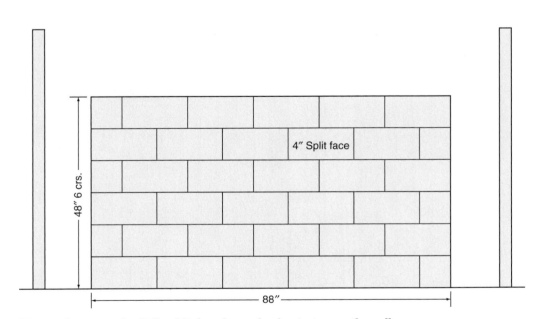

The rough texture of split face block makes a pleasing texture on the wall.

PROCEDURE

1. Items needed:
   supply of shop mortar
   4" × 8" × 16" split face block_____
   4" × 8" × 8" split face half block_____
   When cutting split face to half blocks, use a brick set chisel to "face" cut edge if it is uneven.
2. Mark the corner poles with the proper coursing. Be careful to check the floor or footing for level and make proper adjustments with the first course of block.
3. Mark out bond on the floor or footing.
4. Hang the mason's line on the corner poles at the marks for the first course. *Remember:* If you are using solid units, the bed joint is spread like a brick bed joint. The head joints need to be buttered with care, whether in the wall or before they are laid. Do not drop excess mortar.
5. Lay the first course to the line. *Remember:* Keep $1/16"$ away from the line to avoid pushing out of line (range). Split face block have a rougher edge than smooth block. Place the block as uniformly along the line as possible but do not push the line. As the wall is finished and tooled it will look even and uniform even though the edges of each unit may be rough or inconsistent.
6. After laying the first course, check each jamb end for plumb and alignment with the layout marks.
7. Continue this procedure until you have laid all six courses. Plumb jambs after each course is laid.
8. Tool all joints when thumbprint hard.
9. When the project is done, clean your area and ask the instructor for an evaluation.

Instructor's Evaluation and Checklist

Student's Name: _____

	Possible Points	Points Off
1. Project set up and properly stocked	5	
2. Safety procedures followed	5	
3. Plumb. Check two places. Check range top of third course and top of sixth course.	15	
4. Level. Check top of each course.	10	
5. Height. Check proper course layout and uniform bed joints.	10	
6. Bond. Check for running bond.	10	
7. Tooling: Holes Smears Tags or fingernails that protrude out from wall Uniform shape	15	
8. Proper use of trowel and tools	10	
9. Motion study as described in procedure and good masonry practices	10	
10. Neatness	10	
	100	

TOTAL

FINAL GRADE

_____ SCORE
(Required to pass: Beginning 60–80%, Intermediate 80–90%, Advanced 90%)

Instructor's Comments:

Student Competency #23

TASK

Laying block to the line and building a masonry opening spanned by precast concrete lintels.

PERFORMANCE OBJECTIVES

- The student shall be able to set up and mark corner pole to build an 8″ block wall that is 80″ long and 80″ high with a masonry opening.
- The student shall be able to properly erect and stock scaffold.
- The student shall keep masonry jambs plumb and true, prepare the masonry bearing, and safely handle and set precast lintels in the wall.

RELATED LEARNING ACTIVITIES

1. Properly stock your project according to Competency #2, "Stocking a masonry project."
2. Observe all safety practices. Read and review Unit 29, "Safety Rules for Erecting and Using Scaffolding," of the *Masonry Skills* textbook. Take extra caution when working on scaffold. Use competent help when handling and setting precast lintels.
3. When you begin to work on the scaffold, use a barrier or caution tape to erect a "No Access Zone" on the far side of the wall. It should be 2′ wider than the wall and extend away from your wall 9′. This area should only be entered by you or your instructor, if needed.
4. Read and review Unit 16 of the *Masonry Skills* textbook.
5. Watch the demonstration by the instructor.
6. Review block laying and mortar spreading practices in Competency #5 of this workbook.

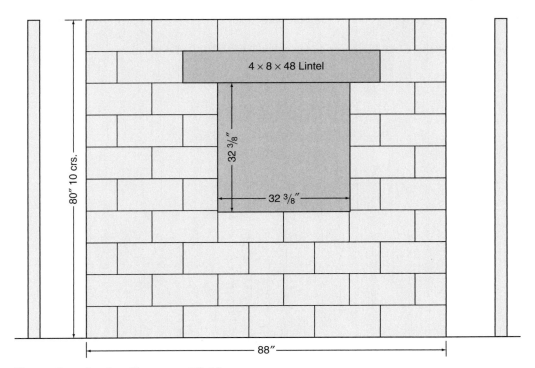

Use caution when handling precast lintels.

PROCEDURE

1. Items needed:
   supply of shop mortar
   8″ block_____
   8″ jamb block_____
   8″ half block
   2 4″ × 8″ × 48″ precast lintels
   masonry tools
2. Mark the corner poles with the proper coursing. Be careful to check the floor or footing for level and make proper adjustments with the first course of block.
3. Mark out bond on the floor or footing.
4. Hang the mason's line on the corner poles at the marks for the first course.
5. Lay the first course to the line. Remember to keep $1/16''$ away from the line to avoid pushing out of line (range).
6. After laying the first course, check each jamb end for plumb and alignment with the layout marks.
7. Continue this procedure until you have four courses laid.
8. Hang the line for the fifth course and mark out the masonry opening according to the project plan.
9. Lay the next four courses on each side of the masonry opening with the line. Remember to plumb all jambs, including the masonry opening.
10. When all block are laid to the top of the masonry opening, fill the core that will be the bearing surface of the lintels with masonry debris or cover with grout screen or hardware cloth.
11. When you lay the ninth course, also lay the far side lintel to the line with the help of a partner. *Remember:* The lintel is marked with a "top." Always keep this side up. Make a clear space on the scaffold for the lintel. Make sure you have a clean, stable path to move in when setting the lintels. After the course is done, have your partner hold the top of the lintel. Put the two-foot level on the underneath of the lintel to check for level. Make any slight adjustment that is needed. Lintels may have a variation so it is important to have true level at the top of the masonry opening. Lay the nearside lintel and check for level underneath and if it is level with the other lintel on the bottom. Do not attempt to fill the space or joint between the backs of the lintel.
12. Lay the last course of block, which will cross over the lintels.
13. Tool all joints when thumbprint hard. Remember to tool under the lintel at the jambs. Use a flat slicker or tip of your trowel.
14. When working on the scaffold, you may raise the walk boards and carefully reach over the wall to tool the other side.
15. When the project is done, clean your area and ask the instructor for an evaluation.

Instructor's Evaluation and Checklist

Student's Name: _____

	Possible Points	Points Off
1. Project set up and properly stocked	15	
2. Safety procedures followed	15	
3. Plumb. Check four places. Range check at seventh and ninth course.	10	
4. Level. Check top fourth course and top. Check under lintels.	10	
5. Height: Check proper course layout and uniform bed joints.	10	
6. Bond. Check for running bond.	5	
7. Tooling: Holes Smears Tags or fingernails that protrude out from wall. Uniform shape	5	
8. Proper use of trowel and tools	10	
9. Motion study as described in procedure and good masonry practices	10	
10. Neatness	10	
	100	

TOTAL

FINAL GRADE

_____ SCORE

(Required to pass: Beginning 60–80%, Intermediate 80–90%, Advanced 90%)

Instructor's Comments:

Student Competency #24

TASK

Laying an 8" corner lead.

Checking the first course on the corner to be sure it is plumb.

PERFORMANCE OBJECTIVE

- The student shall be able to properly lay out and build a six-course-high corner lead with 8" × 8" × 16" block.

RELATED LEARNING ACTIVITIES

1. Properly stock your project according to Competency #2, "Stocking a masonry project."
2. Observe all safety practices.
3. Read and review Unit 17 of the *Masonry Skills* textbook.
4. Watch the demonstration by the instructor.
5. Review Competency #5 of this workbook.

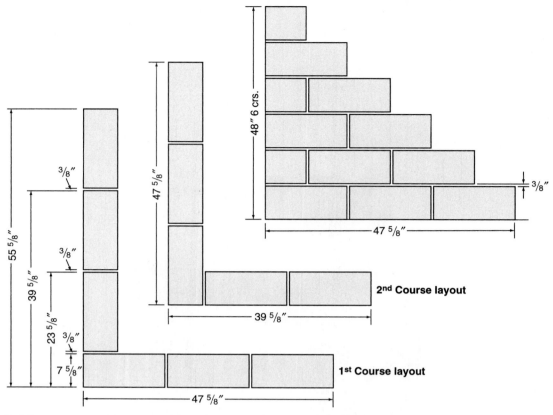

Building a corner lead correctly is important to provide a guide for building the rest of the wall.

Leveling the block. Block should always be leveled by length and width.

Spreading the mortar bed for the first course.

Leveling the first course.

Checking the first course along the top of the block to be sure that it is straight.

Swiping mortar on the block with the trowel. Notice that this technique differs from the one used in brick construction.

Measuring the corner for the proper height. The modular rule should read number two at the top of the corner.

The last step in the construction of a block corner is to check to be certain that the ends of the block are in line.

PROCEDURE

1. Items needed:
   8" block_____
   8" jamb block___6____
   supply of shop mortar
   masonry tools
2. Lay out the project on the floor or foundation according to the plans for this project. Extend wall line layout and bond marks so you can find them after spreading the bed joint.
3. Check the floor or foundation for level. Make sure the first course is level even if the floor or foundation is not. Adjust the thickness of the bed joint if needed.
4. Spread the mortar for the first course in the lead. Lay the first block in the corner and work your way out to the "tail" of the lead. Lay the entire block course as accurately as you can by eye. After the first course is laid, go back and plumb and straightedge each block. Then check for level. After making any adjustments to level the course, go back and recheck plumb and range (straight edge). Make sure you are still on the layout lines and bond marks. Proceed with the second course.
5. Spread the bed joint for the second course. Lay the corner block first. Use the first course as a guide and sight down the corner of the block you are laying just as you set it into final position. Lay all block in that course and then plumb, range, and level them. Recheck before you go onto the next course.
6. Finish the corner in this manner. Do not forget to check for height.
7. Tool joints when thumbprint hard. Use a flat slicker or the tip of your trowel to tool in a nice square inside corner.
8. When you are done with the project, clean your area and ask instructor for an evaluation.

Instructor's Evaluation and Checklist

Student's Name: _____

	Possible Points	Points Off
1. Project set up and properly stocked	5	
2. Safety procedures followed	5	
3. Plumb. Check six places.	10	
4. Level. Check six places.	10	
5. Height. Check six places.	10	
6. Bond: Measure each course lengthwise.	20	

7. Tooling:
  Holes
  Smears
  Tags or fingernails that protrude out from wall
  Uniform shape
	10	

8. Proper use of trowel and tools	10	
9. Motion study as described in project and good masonry practices	10	
10. Neatness	10	
	100		**TOTAL**
			FINAL GRADE

_____ **SCORE**

(Required to pass: Beginning 60–80%, Intermediate 80–90%, Advanced 90%)

Instructor's Comments:

Student Competency #25

TASK

Laying a 10″ corner lead.

PERFORMANCE OBJECTIVE

- The student shall be able to properly lay out and build a six-course-high corner lead with 10″ × 8″ × 16″ block.

RELATED LEARNING ACTIVITIES

1. Properly stock your project according to Competency #2, "Stocking a masonry project."
2. Observe all safety practices.
3. Read and review Unit 17 of the *Masonry Skills* textbook.
4. Watch the demonstration by the instructor.
5. Review Competency #5 of this workbook.
6. The mason must build accurate corner leads that are ranged with the wall line so masons can lay to the line and efficiently and accurately build the wall.

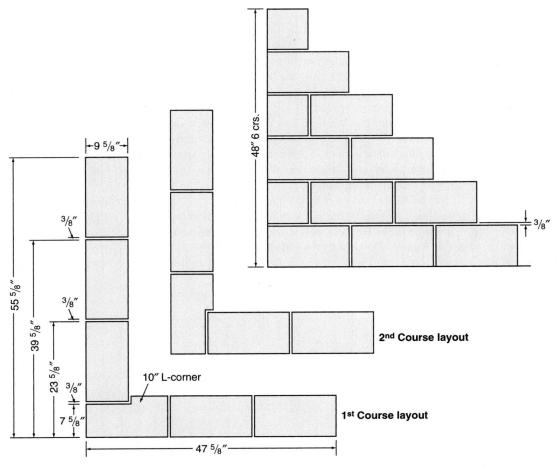

10″ block corner lead.

Student Competency #25

The 10″ L–corner block is used to maintain half lap or running bond.

PROCEDURE

1. Items needed:
   10″ block_____
   10″ L–corner block___6____
   supply of shop mortar
   masonry tools
2. Lay out the project on the floor or foundation according to the plans for this project. Extend wall line layout and bond marks so you can find them after spreading the bed joint.
3. Check the floor or foundation for level. Make sure the first course is level even if the floor or foundation is not. Adjust the thickness of the bed joint if needed.
4. Spread the mortar for the first course in the lead. Lay the first block in the corner and work your way out to the "tail" of the lead. Lay the entire block course as accurately as you can by eye. After the first course is laid, go back and plumb and straightedge each block. Then check for level. After making any adjustments to level the course, go back and recheck plumb and range (straight edge). Make sure you are still on the layout lines and bond marks. Proceed with the second course.
5. Spread the bed joint for the second course. Lay the corner block first. Use the first course as a guide and sight down the corner of the block you are laying just as you set it into final position. Lay all block in that course and then plumb, range, and level them. Recheck before you go on to the next course.
6. Finish the corner in this manner. Do not forget to check for height.
7. Tool joints when thumbprint hard. Use a flat slicker or the tip of your trowel to tool in a nice square inside corner.
8. When you are done, clean your area and ask instructor for an evaluation.

Instructor's Evaluation and Checklist

Student's Name: _____

	Possible Points	Points Off
1. Project set up and properly stocked	5	
2. Safety procedures followed	5	
3. Plumb. Check six places.	10	
4. Level. Check six places.	10	
5. Height. Check six places.	10	
6. Bond. Measure each course lengthwise.	20	
7. Tooling: 　Holes 　Smears 　Tags or fingernails that protrude out from wall 　Uniform shape	10	
8. Proper use of trowel and tools	10	
9. Motion study as described in project and good masonry practices	10	
10. Neatness	10	
	100	

TOTAL

FINAL GRADE

_____ SCORE

(Required to pass: Beginning 60–80%, Intermediate 80–90%, Advanced 90%)

Instructor's Comments:

Student Competency #26

TASK

Laying a 12' corner lead.

PERFORMANCE OBJECTIVE

- The student shall be able to properly lay out and build a six-course-high corner lead with 12" × 8" × 16" block.

RELATED LEARNING ACTIVITIES

1. Properly stock your project according to Competency #2, "Stocking a masonry project."
2. Observe all safety practices.
3. Read and review Unit 17 of the *Masonry Skills* textbook.
4. Watch the demonstration by the instructor.
5. Review Competency #5 of this workbook.

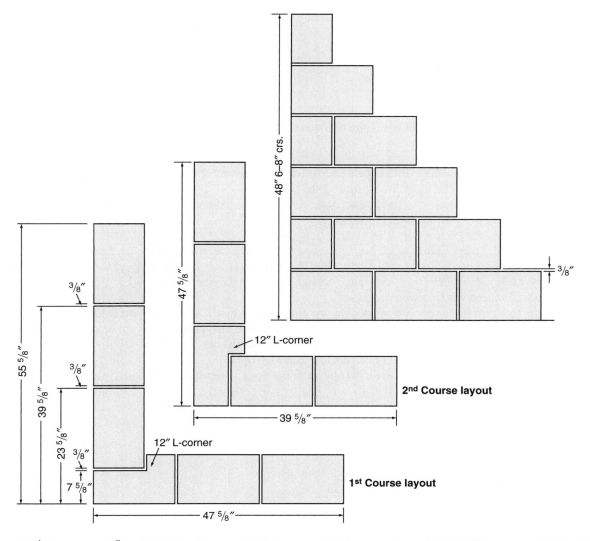

A 12' L–corner or 8" jamb block and concrete brick are used in the corner to maintain half lap or running bond.

108 Student Competency #26

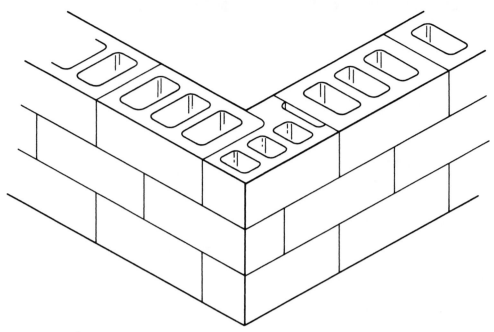

L–shaped, 12" corner block laid in position.

PROCEDURE

1. Items needed:
   12" block_____
   12" L–corner block___6____ OR 6 8" jamb block and 6 concrete brick
   supply of shop mortar
   masonry tools
2. Lay out the project on the floor or foundation according to the plans for this project. Extend wall line layout and bond marks so you can find them after spreading the bed joint.
3. Check the floor or foundation for level. Make sure the first course is level even if the floor or foundation is not. Adjust the thickness of the bed joint if needed.
4. Spread the mortar for the first course in the lead. Lay the first block in the corner and work your way out to the "tail" of the lead. Lay the entire block course as accurately as you can by eye. After the first course is laid, go back and plumb and straightedge each block. Then check for level. After making any adjustments to level the course, go back and recheck plumb and range (straight edge). Make sure you are still on the layout lines and bond marks. Proceed with the second course.
5. Spread the bed joint for the second course. Lay the corner block first. Use the first course as a guide and sight down the corner of the block you are laying just as you set it into final position. Lay all block in that course and then plumb, range, and level them. Recheck before you go onto the next course.
6. Finish the corner in this manner. Do not forget to check for height.
7. Tool joints when thumbprint hard. Use a flat slicker or the tip of your trowel to tool in a nice square inside corner.
8. When you are done with your project, clean your area and ask instructor for an evaluation.

Instructor's Evaluation and Checklist

Student's Name: _____

	Possible Points	Points Off
1. Project set up and properly stocked	5	
2. Safety procedures followed	5	
3. Plumb. Check six places.	15	
4. Level. Check six places.	15	
5. Height. Check six places.	10	
6. Bond. Measure each course lengthwise.	10	

7. Tooling:
    Holes
    Smears
    Tags or fingernails that protrude out from wall
    Uniform shape
	10	

8. Proper use of trowel and tools	10	
9. Motion study as described in project and good masonry practices.	10	
10. Neatness	10	
	100		TOTAL
			FINAL GRADE

_____ SCORE

(Required to pass: Beginning 60–80%, Intermediate 80–90%, Advanced 90%)

Instructor's Comments:

Student Competency #27

TASK

Laying a 6" corner lead.

PERFORMANCE OBJECTIVE

■ The student shall be able to properly lay out and build a six-course-high corner lead with 6" × 8" × 16" block.

RELATED LEARNING ACTIVITIES

1. Properly stock your project according to Competency #2, "Stocking a masonry project."
2. Observe all safety practices.
3. Read and review Unit 17 of the *Masonry Skills* textbook.
4. Watch the demonstration by the instructor.
5. Review Competency #5 of this workbook.

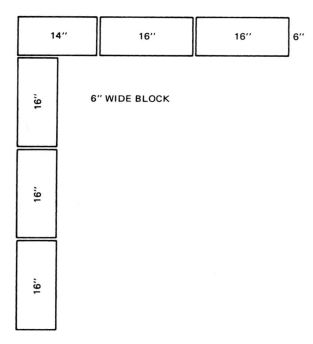

Corner layout for 6" lead.

PROCEDURE

1. Items needed:
   6" block_____
   13⁵⁄₈" cuts for corner block___6___
   supply of shop mortar
   masonry tools
2. Lay out the project on the floor or foundation according to the plans for this project. Extend wall line layout and bond marks so you can find them after spreading the bed joint.
3. Check the floor or foundation for level. Make sure the first course is level even if the floor or foundation is not. Adjust the thickness of the bed joint if needed.

111

4. Spread the mortar for the first course in the lead. Lay the first block in the corner and work your way out to the "tail" of the lead. Lay the entire block course as accurately as you can by eye. After the first course is laid, go back and plumb and straightedge each block. Then check for level. After making any adjustments to level the course, go back and recheck plumb and range (straight edge). Make sure you are still on the layout lines and bond marks. Proceed with the second course.
5. Spread the bed joint for the second course. Lay the corner block first. Use the first course as a guide and sight down the corner of the block you are laying just as you set it into final position. Lay all block in that course and then plumb, range, and level them. Recheck before you go on to the next course.
6. Finish the corner in this manner. Do not forget to check for height.
7. Tool joints when thumbprint hard. Use a flat slicker or the tip of your trowel to tool in a nice square inside corner.
8. When you are done with your project, clean your area and ask instructor for an evaluation.

Student Competency #27

Instructor's Evaluation and Checklist

Student's Name: _____

	Possible Points	Points Off
1. Project set up and properly stocked	5	
2. Safety procedures followed	5	
3. Plumb. Check six places.	15	
4. Level. Check six places.	15	
5. Height. Check six places.	10	
6. Bond. Measure each course lengthwise.	10	

7. Tooling:
   Holes
   Smears
   Tags or fingernails that protrude out from wall
   Uniform shape
	10	

8. Proper use of trowel and tools	10	
9. Motion study as described in project and good masonry practices	10	
10. Neatness	10	
	100		**TOTAL**

FINAL GRADE

_____ **SCORE**

(Required to pass: Beginning 60–80%, Intermediate 80–90%, Advanced 90%)

Instructor's Comments:

Student Competency #28

TASK

Laying a 4″ corner lead.

Student mason laying out 4″ block corner with a 12″ cut block.

PERFORMANCE OBJECTIVE

■ The student shall be able to properly lay out and build a six-course-high corner lead with 4″ × 8″ × 16″ block.

RELATED LEARNING ACTIVITIES

1. Properly stock your project according to Competency #2, "Stocking a masonry project."
2. Observe all safety practices.
3. Read and review Unit 17 of the *Masonry Skills* textbook.
4. Watch the demonstration by the instructor.
5. Review Competency #5 of this workbook.

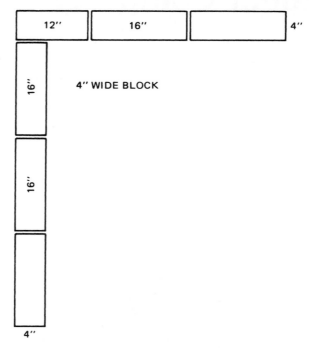

Corner layout for 4″ block lead.

PROCEDURE

1. Items needed:
   4″ block_____
   11⅝″ cuts for corner block___6____
   supply of shop mortar
   masonry tools
2. Lay out the project on the floor or foundation according to the plans for this project. Extend wall line layout and bond marks so you can find them after spreading the bed joint.
3. Check the floor or foundation for level. Make sure the first course is level even if the floor or foundation is not. Adjust the thickness of the bed joint if needed.
4. Spread the mortar for the first course in the lead. Lay the first block in the corner and work your way out to the "tail" of the lead. Lay the entire block course as accurately as you can by eye. After the first course is laid, go back and plumb and straightedge each block. Then check for level. After making any adjustments to level the course, go back and recheck plumb and range (straight edge). Make sure you are still on the layout lines and bond marks. Proceed with the second course.
5. Spread the bed joint for the second course. Some 4″ blocks have aggregate in the top of the cores so you may spread the bed joint as for brick. Lay the corner block first. Use caution, because 4″ block tend to tip easily. Use the first course as a guide and sight down the corner of the block you are laying just as you set it into final position. Lay all block in that course and then plumb, range, and level them. Recheck before you go on to the next course.
6. Finish the corner in this manner. Do not forget to check for height.
7. Tool joints when thumbprint hard. Use a flat slicker or the tip of your trowel to tool in a nice square inside corner.
8. When you are done with the project, clean your area and ask instructor for an evaluation.

Instructor's Evaluation and Checklist

Student's Name: _____

	Possible Points	Points Off
1. Project set up and properly stocked	5	
2. Safety procedures followed	5	
3. Plumb. Check six places.	15	
4. Level. Check six places.	15	
5. Height. Check six places.	10	
6. Bond. Measure each course lengthwise.	10	
7. Tooling: Holes Smears Tags or fingernails that protrude out from wall Uniform shape	10	
8. Proper use of trowel and tools	10	
9. Motion study as described in project and good masonry practices	10	
10. Neatness	10	
	100	

TOTAL

FINAL GRADE

_____ SCORE

(Required to pass: Beginning 60–80%, Intermediate 80–90%, Advanced 90%)

Instructor's Comments:

Student Competency #29

TASK

Building a block pier.

PERFORMANCE OBJECTIVES

- The student shall be able to properly lay out and build an 8″ block pier 88″ high.
- The student shall be able to maintain plumb, level, coursing, square, and range on all four sides of the pier.

RELATED LEARNING ACTIVITIES

1. Properly stock your project according to Competency #2, "Stocking a masonry project."
2. Observe all safety practices. Read and review Unit 29 of *Masonry Skills* textbook.
3. When your project is six courses high, erect the scaffold. Mark off a "No Access Zone" around the pier. This area is restricted to everyone but you or the instructor.
4. Review Unit 17 of *Masonry Skills* textbook. The techniques for building a block pier are similar to building a block corner because it is all level work.
5. Watch the demonstration by the instructor.

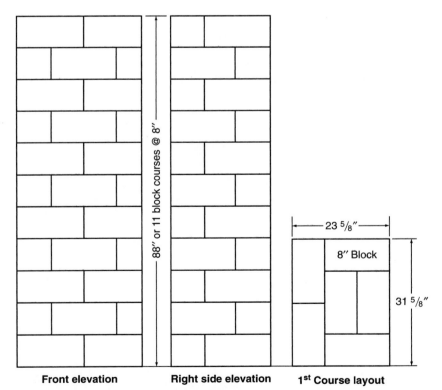

Front elevation   **Right side elevation**   **1st Course layout**

It takes practice and skill to keep a pier plumb the entire height.

PROCEDURE

1. Items needed:
   8" jamb block_____
   supply of shop mortar
   masonry tools
2. Layout the project on floor or foundation according to project plans.
3. Lay the first course. Check for plumb square level and range. Make sure you have the proper dimensions and height, which should be 8" plus or minus depending on the levelness of the floor or foundation.
4. Lay the next five courses. Be careful not to drop excess mortar in the cavity of the pier. Check for height. It should be in 8" increments or #2 on the modular scale measuring from the top of the first course.
5. After completing seven courses, build and stock the scaffold.
6. When you lay the pier from the scaffold, lay the blocks on the far side of the pier first. You can lean over to plumb and level them without the nearside course being in the way.
7. Tool joints when thumbprint hard.
8. When project is done, clean your area and ask instructor for evaluation.

Instructor's Evaluation and Checklist

Student's Name: _____

	Possible Points	Points Off
1. Project set up and properly stocked	15	
2. Safety procedures followed	15	
3. Plumb. Check six places full height.	10	
4. Level. Check third and seventh joint. Check top.	5	
5. Height. Check modular #2 spacing or 8' coursing.	10	
6. Bond. Check for uniform appearance when viewing from a distance of 5 feet.	5	
7. Tooling: Holes Smears Tags or fingernails that protrude out from the face of the wall Uniform shape	10	
8. Proper use of trowel and tools	10	
9. Motion study as described in procedure and good masonry practices	10	
10. Neatness	10	
	100	

TOTAL

FINAL GRADE

_____ SCORE

(Required to pass: Beginning 60–80%, Intermediate 80–90%, Advanced 90%)

Instructor's Comments:

SECTION SIX
ESTIMATING BRICK AND CONCRETE BLOCK BY RULE OF THUMB MATH AND CUTTING WITH THE MASONRY SAW

SECTION SIX
ESTIMATING BRICK AND CONCRETE BLOCK QUANTITIES OF MATERIALS AND CUTTING WITH THE MASONRY SAW

Student Competency #30

TASK

Cutting masonry units with the saw.

Holding a thin cut with waste pieces to prevent getting the hand too close to the blade.

PERFORMANCE OBJECTIVES

- The student shall be able to cut block and brick to commonly used shapes and dimensions.
- The student shall know how to properly control dust produced by cutting with the masonry saw.

RELATED LEARNING ACTIVITIES

1. Observe all safety practices. *Safety glasses, ear protection, and dust protection are required.* Do not have loose-fitting sleeves, dangling jewelry, or hanging long hair that could be caught in saw.
2. Cut wet to control dust. If you cannot cut wet, use proper exhaust fans or proper respirator to eliminate dust hazards.
3. Watch the demonstration by the instructor.
4. Read and review Unit 21 of the *Masonry Skills* textbook.
5. The brick cut with a 45-degree angle and four $1^1/_2''$ deep slots can be cleaned and sealed to use as a letter or note holder.

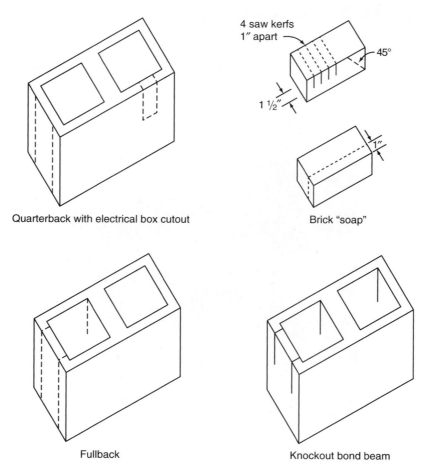

Quarterback with electrical box cutout

Brick "soap"

Fullback

Knockout bond beam

The brick with the 45-degree cut and four $1/2''$ slots can be cleaned and used as a letter holder.

Using the miter gauge to hold brick securely when cutting angle pieces.

Mason cutting a rip block on the saw (wet saw with a diamond blade). Notice the proper clothing and eye protection.

Fullback cut block. Notice that all the webs have been cut but the last one.

Electric receptacle box cut into a concrete block corner. Notice the entire box is enclosed in the wall.

PROCEDURE

1. Items needed:
   3 concrete block
   2 common brick
   masonry tools
2. Check saw to see if blade is properly attached. If using a handheld chop saw, check fuel. *Remember:* Only cut outdoors or in properly vented space if using gas-powered saw.
3. Cut the three blocks as shown in the project plan sheet. Cut the electrical box first.
4. To cut the electrical box opening, trace around the box with a pencil on the block.
5. Set the block on saw table and cut successive $1/2''$ cuts from top of block down to bottom of box location. Carefully chip out with blade of the brick hammer. Striking from the face of the block may crack it.
6. Cut the brick as shown on the project plan. To cut the brick "soap," use waste pieces to hold it so your fingers are not too close to the blade.

Bond-beam block cut out of a full block. Notice that the lower part of the block is left in for the placement of the steel rods and concrete.

Instructor's Evaluation and Checklist

Student's Name: _____

	Possible Points	Points Off
1. Safety procedures followed. Dust mask, eye protection, ear protection worn.	15	
2. Checked saw is in proper working order.	10	
3. Halfback with electrical box cut out	15	
4. Fullback	15	
5. Knockout bond beam	15	
6. Brick with 45-degree angle and saw kerfs to use as letter holder	15	
7. Brick cut to "soap" piece	15	
	100	

TOTAL

FINAL GRADE

_____ SCORE

(Required to pass: Beginning 60–80%, Intermediate 80–90%, Advanced 90%)

Instructor's Comments:

SECTION SEVEN
MASONRY PRACTICES AND DETAILS OF CONSTRUCTION

Student Competency #31

TASK

Building a brick pier.

PERFORMANCE OBJECTIVES

- The student shall be able to properly lay out and build a brick pier with block backup.
- The student shall be able to maintain plumb, level, coursing, and square on all four sides.

RELATED LEARNING ACTIVITIES

1. Properly stock your project as described in workbook Competency #2.
2. Observe all safety practices.
3. Read and review Unit 11 of *Masonry Skills* textbook. The techniques for building a brick pier are very similar to building a brick corner because it is all level work.
4. Watch the demonstration by the instructor.

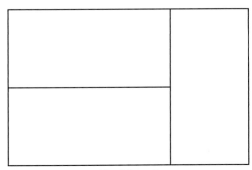

Block layout

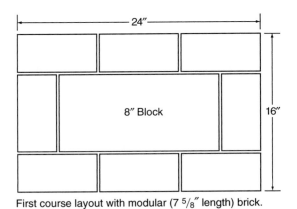

First course layout with modular (7 5/8" length) brick.

This pier has one-course-high block foundation.

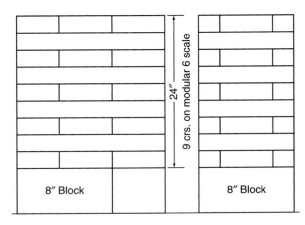

Front and side elevation of pier.

PROCEDURE

1. Items needed:
   8" jamb block_____
   brick_____
   supply of shop mortar
   6 wall ties
   masonry tools
2. Lay out project on floor or foundation according to plan in workbook.
3. Lay the base with three 8" jamb block. Recheck dimensions, square, and level before continuing. Fill the cores with masonry debris or cover with grout screen or hardware cloth to make solid bearing for the brick.
4. Lay the first course of brick in the pier. Check plumb, level, range, and coursing.
5. Continue building the pier by laying one of the 24" walls three courses high before doing the other side. As you lay the 24" side you will have to lay the return brick (tail of lead) also. When this "lead" is three courses high, fill in the courses on the other side of the pier. After you reach six total courses of brick, lay in the two courses of 8" block. Install wall ties and continue with the last three courses.
6. Tool joints when thumbprint hard.
7. When you are done with the project, clean your area and ask the instructor for an evaluation.

Instructor's Evaluation and Checklist

Student's Name: _____

	Possible Points	Points Off
1. Project set up and properly stocked	5	
2. Safety procedures followed	5	
3. Plumb. Check six places.	15	
4. Level. Check four places.	10	
5. Height. Check modular #6 spacing.	15	
6. Bond. Check for uniform joint appearance when viewing from a distance of 5 feet.	10	
7. Tooling: Holes Smears Tags or fingernails that protrude from wall Uniform shape	10	
8. Proper use of trowel and tools	10	
9. Motion study as described in procedure and good masonry practices	10	
10. Neatness	10	
	100	

TOTAL

FINAL GRADE

_____ SCORE

(Required to pass: Beginning 60–80%, Intermediate 80–90%, Advanced 90%)

Instructor's Comments:

Student Competency #32

TASK

Building a six-corner brick pier.

PERFORMANCE OBJECTIVES

- The student shall be able to properly lay out and build a brick pier with block backup.
- The student shall be able to maintain plumb, level, coursing, and square on all six sides.

RELATED LEARNING ACTIVITIES

1. Properly stock your project as described in workbook Competency #2.
2. Observe all safety practices.
3. Read and review Unit 11 of *Masonry Skills* textbook. The techniques for building a brick pier are very similar to building a brick corner because it is all level work.
4. Watch the demonstration by the instructor.

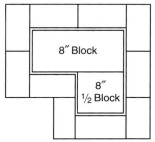

2nd Course layout of brick

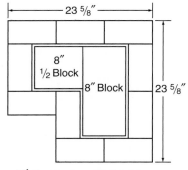

1st Course layout of brick

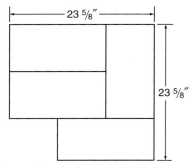

Layout for first course of block

Plan view for six-corner brick pier.

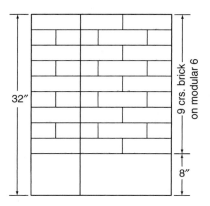

PROCEDURE

1. Items needed:
   8" jamb block_____
   brick_____
   supply of shop mortar
   masonry tools
2. Lay out project on floor or foundation according to plan in workbook.
3. Lay the base with four 8" jamb block. Recheck dimensions, square, and level before continuing. Fill the cores with masonry debris or cover with grout screen or hardware cloth to make solid bearing for the brick.
4. Lay the first course of brick in the pier. Check plumb, level, range, and coursing.
5. Continue building the pier by laying one of the 24" walls three courses high before doing the other sides. As you lay the 24" side you will have to lay the return brick (tail of lead) also. When this "lead" is three courses high, fill in the courses on the other sides of the pier. Use this lead to level and align your other courses instead of measuring. After you reach six courses, lay in the two courses of 8" block. Install wall ties and continue with the last three courses.
6. Tool joints when thumbprint hard.
7. When you are done, clean your area and ask the instructor for an evaluation.

Instructor's Evaluation and Checklist

Student's Name: _____

	Possible Points	Points Off
1. Project set up and properly stocked	5	
2. Safety procedures followed	5	
3. Plumb. Check six places.	15	
4. Level. Check four places.	10	
5. Height. Check modular #6 spacing.	15	
6. Bond. Check for uniform joint appearance when viewing from a distance of 5 feet.	10	

7. Tooling:
    Holes
    Smears
    Tags or fingernails that protrude from wall
    Uniform shape
	10	

8. Proper use of trowel and tools	10	
9. Motion study as described in procedure and good masonry practices	10	
10. Neatness	10	
	100	

TOTAL

FINAL GRADE

_____ SCORE

(Required to pass: Beginning 60–80%, Intermediate 80–90%, Advanced 90%)

Instructor's Comments:

Student Competency #33

TASK

Building a composite wall.

Prefabricated metal joint reinforcement extending past walls could cause serious injury.

PERFORMANCE OBJECTIVES

- The student shall properly lay out and build a 12" wide composite wall using horizontal joint reinforcement to tie the wall together every 16" of height.
- The student shall maintain plumb, level, range, dimension, and coursing on both wythes of masonry.

RELATED LEARNING ACTIVITIES

1. Properly stock your project as described in workbook Competency #2.
2. Observe all safety practices. Use caution when handling joint reinforcement.
3. Read and review Unit 22 of the *Masonry Skills* textbook.
4. Watch the demonstration by instructor.

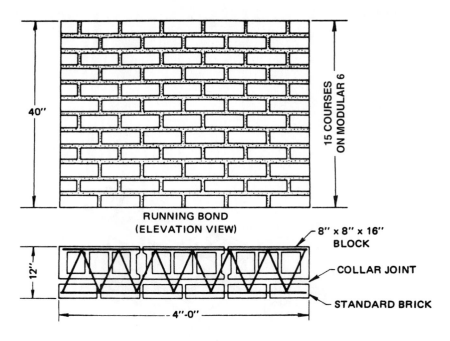

Laying a 12″ brick and concrete block composite panel wall bonded with masonry wire reinforcement.

PROCEDURE

1. Items needed:
   supply of shop mortar
   masonry tools
   8″ jamb block_____
   8″ half block
   8″ block
   brick
   12″ horizontal joint reinforcement
2. Lay out project lines on floor or foundation according to plans. Check floor for level and make adjustment in first course of block if needed.
3. Lay the first two courses of block. Be sure to stay on 8″ coursing or #2 on modular scale. For this project *do not* fill the collar joint or parge the backup.
4. Mark out the brick coursing using the #6 modular scale. Make sure you come up even with the top of the second course of block that is level.
5. Lay out brick bond and then build six-course lead at one end. Remember to keep the brick jamb flush with the block jamb. Do not let excess mud build up in collar joint. It may push your wall apart.
6. Lay the brick lead at the other end. You can use the first lead and your level as a guide for height and range.
7. After you finish six courses, install the horizontal joint reinforcement and lay the next two courses of block.
8. Mark out the brick courses and lay the brick lead on one end as before. Finish the project in this manner.
9. Tool joints when thumbprint hard.
10. When you are done with the project, clean your area and ask your instructor for an evaluation.

Instructor's Evaluation and Checklist

Student's Name: _____

	Possible Points	Points Off
1. Project set up and properly stocked	5	
2. Safety procedures followed	5	
3. Plumb. Check ten places.	15	
4. Level. Check six places.	15	
5. Height. Check four places.	10	
6. Bond. Check plumb bond in two places.	10	
7. Tooling: Holes Smears Fingernails and tags that protrude from the face of the wall Uniform shape	10	
8. Proper use of trowel and tools	10	
9. Motion study according to project plans and good masonry practices	10	
10. Neatness	10	
	100	

TOTAL

FINAL GRADE

_____ SCORE

(Required to pass: Beginning 60–80%, Intermediate 80–90%, Advanced 90%)

Instructor's Comments:

Student Competency #34

TASK

Brick on horizontal relief angle.

PERFORMANCE OBJECTIVE

- The student will be able to build a brick wall to proper height; adjust, set, and level a brick relief angle; and properly install flashing, brick ties, and weep holes.

RELATED LEARNING ACTIVITIES

1. Properly stock your project as described in workbook Competency #2.
2. Observe all safety practices.
3. The school or training facility should have installed three $3/4''$ all thread bolts in a permanent masonry wall. Refer to project plans for layout.
4. Read and review Unit 25 of *Masonry Skills* textbook.
5. On masonry jobs using relief angles, notice they are installed "loose" by other trades. It is up to the mason to adjust and set into final position. Do not remove relief angle from project wall. Adjust it to highest position possible and tighten bolts before starting project.

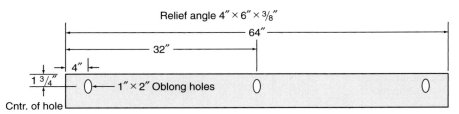

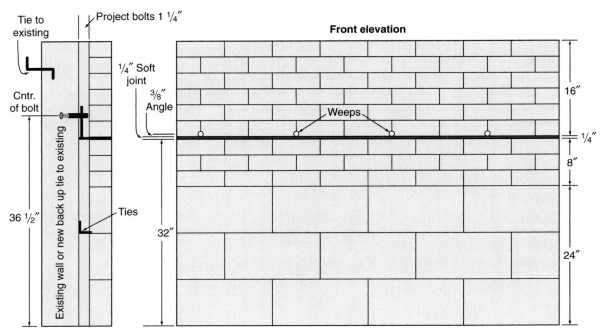

Details of relief angle.

146 Student Competency #34

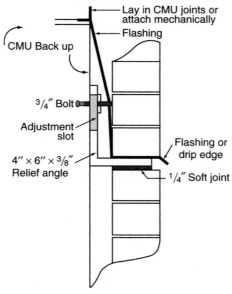

Horizontal expansion joints allow for brick expansion or movement from dissimilar materials. Masonry work must be level and uniform to receive angle.

PROCEDURE

1. Items needed:
   4″ block_____
   brick_____
   supply of shop mortar
   masonry tools
   $1/4″$ premolded brick joint or "soft joint"
   cavity wall flashing
2. Study plans and verify coursing marks on wall. Loosen relief angle and raise it to highest position and tighten bolts so it does not slide down. *Do not remove from wall.*
3. Build masonry project to proper height to receive relief angle as specified. Build leads at each end and hang line to fill in.
4. Tool Joints when thumbprint hard.
5. Install $1/4″$ soft-joint on top course, keeping back $1/2″$ from face of brick.
6. Loosen relief angle and carefully set it down on soft joint. Check for level and tighten bolts. If angle springs up out of level or twist after tightening bolts, you will have to put shims behind angle and retighten.
7. Install flashing so that it projects $1/2″$ past face of exterior wall. Keep flashing smooth and tucked back against angle as best you can. Do not puncture flashing on bolts. If metal drip edge is provided, lap cavity wall flashing on top of metal drip edge and back from face of wall 1″.
8. Dry lay the first course of brick on the relief angle. Check bond with work below, and if brick doesn't fit by bolt heads, cut the back out of those brick that will hit bolts. When you are sure everything fits, continue with the last six courses.
9. Lay the last six courses of brickwork on the flashing, making sure it is plumb with the brick below. Install weep vents every two bricks. Weeps should sit directly on flashing and not have mortar under them. Keep cavity clean.
10. When you are done with your project, clean your area and ask instructor for an evaluation.

Student Competency #34

Instructor's Evaluation and Checklist

Student's Name: _____

	Possible Points	Points Off
1. Project set up and properly stocked	10	
2. Safety procedures followed	5	
3. Plumb. Check five places.	15	
4. Level. Check two places. Bricks in course under relief angle must be level.	10	
5. Height. Check coursing in two places.	5	
6. Bond. Check plumb bond in two places.	5	
7. Tooling: Expansion joint mortar free Holes Smears Tags or fingernails that protrude from face of wall Uniform shape	15	
8. Proper use of trowel and tools.	5	
9. Motion study according to project plans and good masonry practices. Relief angle was not removed from wall.	15	
10. Flashing and weep vents installed properly	15	
	100	

TOTAL

FINAL GRADE

_____ SCORE
(Required to pass: Beginning 60–80%, Intermediate 80–90%, Advanced 90%)

Instructor's Comments:

Student Competency #35

TASK

Building a cavity wall.

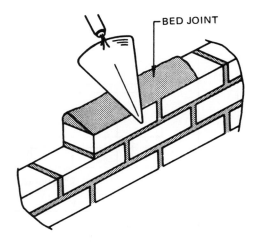

Beveling the bed joint with the blade of the trowel.

PERFORMANCE OBJECTIVES

- The student shall properly lay out and build a 14" wide cavity wall using hook and eye horizontal joint reinforcement to tie the wall together every 16" of height.
- The student shall maintain plumb, level, range, dimension, and coursing on both wythes of masonry.

RELATED LEARNING ACTIVITIES

1. Properly stock your project as described in workbook Competency #2.
2. Observe all safety practices. Use caution when handling joint reinforcement.
3. Read and review Unit 23 of the *Masonry Skills* textbook.
4. Watch the demonstration by instructor.

PROCEDURE

1. Items needed:
    supply of shop mortar
    masonry tools
    8" jamb block_____
    8" half block
    8" block
    brick
    8" horizontal joint reinforcement with eyes to receive ties wire ties that hook into joint reinforcement cavity wall or through wall flashing (scrap rubber roofing or six mil polyethylene sheeting can be substituted)
2. Lay out project lines on floor or foundation according to plans. Check floor for level and make adjustment in first course of block if needed.

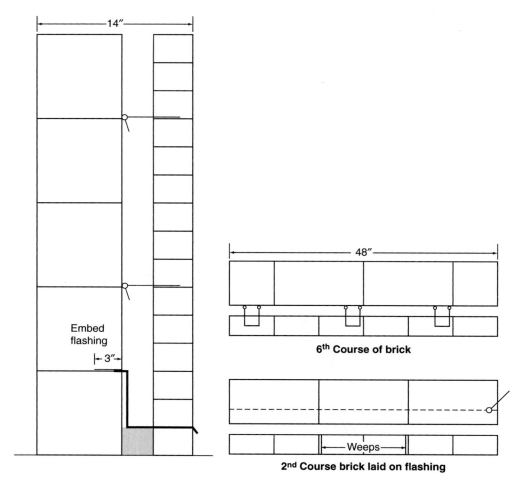

Cavity wall.

3. Lay the first course of block. Double-check plumb, level, dimension, and range before you continue. Be sure to stay on 8″ coursing or #2 on modular scale. Install the through wall flashing on top of the first course of block as shown on the project plan. Spread the bed joint and lay the second course of block, and then install the horizontal joint reinforcement. Continue to lay block to the top of the project. Do not forget to install horizontal joint reinforcement every 16″.
4. Mark out the brick coursing on the block using the #6 modular scale. Make sure the brick coursing comes out even with the block.
5. Lay out brick bond and then lay the first course of brick. Carefully fill the cavity between the block and the first course of brick with mortar to the top of the brick so the flashing will not sag. Before continuing with brickwork, lap the flashing down the face of the block and out over the first course of brick. Keep flashing smooth and even. Let it project past face of brick $1/2″$.
6. Lay a six-course lead at one end. Leave weep holes in every other head joint of the course on top of the flashing. Weep holes may be head joints left out or cotton mop rope laid on the flashing in the head joint. Remember to keep the brick jamb flush with the block jamb. Do not let excess mud build up in the cavity. Bevel the far side of the bed joint slightly to keep excess mud from squeezing out into the cavity when you lay the brick.
7. Lay the brick lead at the other end. You can use the first lead and your level as a guide for height and range.
8. After you finish six courses, install the brick ties and lay the next six courses of brick.
9. Finish the project in this manner.
10. Tool joints when thumbprint hard.
11. When you are done with the project, clean your area and ask your instructor for an evaluation.

Instructor's Evaluation and Checklist

Student's Name: _____

	Possible Points	Points Off
1. Project set up and properly stocked	5	
2. Safety procedures followed	5	
3. Plumb. Check ten places.	15	
4. Level. Check six places.	10	
5. Height. Check four places.	10	
6. Bond. Check plumb bond in two places. Ties installed properly.	10	
7. Tooling: Holes Smears Fingernails and tags that protrude from the face of the wall Uniform shape	10	
8. Proper use of trowel and tools Mortar in cavity? Mortar bridge to block?	15	
9. Motion study according to project plans and good masonry practices. Flashing installed properly	10	
10. Neatness	10	
	100	

TOTAL

FINAL GRADE

_____ SCORE

(Required to pass: Beginning 60–80%, Intermediate 80–90%, Advanced 90%)

Instructor's Comments:

Student Competency #36

TASK

Building movement joints and bond beam in a composite wall.

Wide flange rapid control joint.

PERFORMANCE OBJECTIVE

■ The student shall properly lay out and build a 12" wide composite wall. The wall shall include a control joint in the block wall, an expansion joint in the brick wall, and a bond beam on the last course of block.

RELATED LEARNING ACTIVITIES

1. Properly stock your project as described in workbook Competency #2.
2. Observe all safety practices. Use caution when handling joint reinforcement and cutting rebar.
3. Read and review Unit 22 of the *Masonry Skills* textbook.
4. Watch the demonstration by instructor.
5. This project requires you to cut, install, and position rebar. Do not grout the bond beam course in this project

154 Student Competency #36

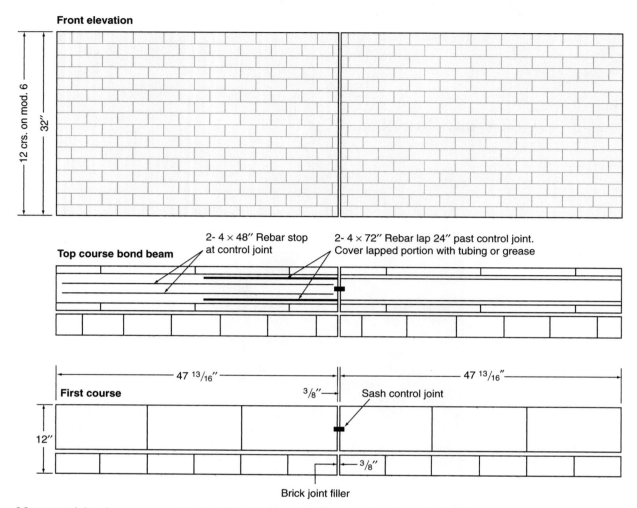

Movement joints in masonry walls should be plumb and uniform so they do not detract from the appearance.

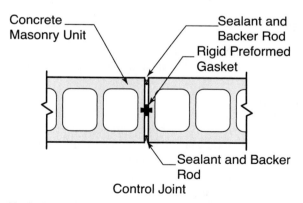

Typical control joint in a concrete masonry unit wall.
(Courtesy of Brick Industry Association)

PROCEDURE

1. Items needed:
   supply of shop mortar
   masonry tools
   8″ sash/jamb block_____
   8″ sash/half block
   8″ block
   brick
   12″ horizontal joint reinforcement
   preformed sash control joint gasket
   premolded compressible $3/8″$ joint filler for brick
   No. 4 ($1/2″$) rebar
   2 24″ pieces $5/8″$ of plastic water line
   grout screen
2. Lay out project lines and bond on floor or foundation according to plans. Check floor for level and make adjustment in first course of block if needed.
3. Build a two-course lead at each end. Be sure to stay on 8″ coursing or #2 on modular scale. For this project *do not* fill the collar joint or parge the back up.
4. Hang the line and lay the first two courses of block. Make sure the control joint stays plumb.
5. Mark out the brick coursing using the #6 modular scale. Make sure you come up even with the top of the second course of block that is level.
6. Lay out brick bond and then build a six-course lead at one end. Remember to keep the brick jamb flush with the block jamb. Do not let excess mud build up in collar joint. It may push your wall apart.
7. Lay the brick lead at the other end then hang the line and lay the six courses. Remember to keep the expansion joint of the brick plumb and in the same position as the block control joint.
8. After you finish six courses, install the horizontal joint reinforcement and lay the next two courses of block. Remember, do not continue the joint reinforcement through the movement joints.
9. Mark out the brick courses and lay the brick lead on one end as before. Finish the project in this manner. Remember the last course of block is a bond beam.
10. If you do not have factory bond beam or knock out bond beam, cut the proper amount of block into knock out bond beam. Use or make a sash end to continue control joint gasket through bond beam. Lay the bond beam course. If you are using open bottom bond beam, roll out grout screen or hardware cloth before spreading the bed joint for the bond beam course.
11. Tool joints when thumbprint hard.
12. Install rebar in the bond beam as shown on the project plan. Make sure you have properly used the slip joint so that there is no bonded or restrained rebar continuing through the control joint.
13. When you are done with the project, clean your area and ask your instructor for an evaluation.

Student Competency #36

Instructor's Evaluation and Checklist

Student's Name: _____

	Possible Points	Points Off
1. Project set up and properly stocked	5	
2. Safety procedures followed	5	
3. Plumb. Check ten places.	15	
4. Level. Check six places.	15	
5. Height. Check four places.	10	
6. Bond. Check plumb bond in two places.	10	
7. Tooling: Holes Smears Fingernails and tags that protrude from the face of the wall Uniform shape	10	
8. Proper use of trowel and tools	10	
9. Motion study according to project plans and good masonry practices	10	
10. Neatness	10	
	100	

TOTAL

FINAL GRADE

_____ SCORE
(Required to pass: Beginning 60–80%, Intermediate 80–90%, Advanced 90%)

Instructor's Comments:

Student Competency #37

TASK

Building pilasters and a masonry opening in a block wall.

PERFORMANCE OBJECTIVE

- The student shall able to properly lay out and build a masonry wall with bonded pilasters and a masonry opening spanned by built-in-place masonry lintel.

RELATED LEARNING ACTIVITIES

1. Properly stock your project as described in workbook Competency #2.
2. Observe all safety practices.
3. Sometimes masons build cast-in-place lintels to span a masonry opening. Bond beam or lintel block are laid on a temporary support, rebar is installed, and the grout is poured to the top of the bond beam or lintel block. The form is removed after 7 or more days of curing.
4. Read and review Unit 24 of the *Masonry Skills* textbook.
5. Watch the demonstration by the instructor.

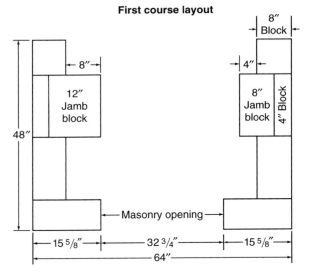

First course layout of project 35. The pilasters will be bonded by every other block course.

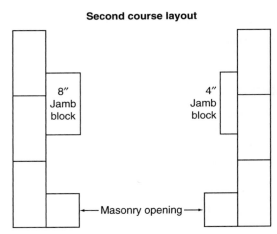

Second course layout of project 35. Do not fill collar joint in pilaster. This would push block out of plumb.

Student Competency #37

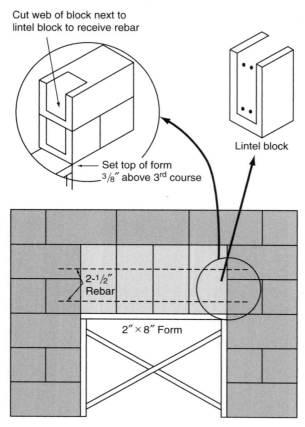

Front view of project 35. Be sure form is secure and $3/8''$ above masonry jamb.

Project ready to lay lintel block and last course. Make sure form is level length ways and across width.

PROCEDURE

1. Items needed:
   8″ jamb block_____
   8″ half block_____
   8″ block _____
   8″ lintel block_____
   12″ jamb block_____
   4″ jamb block._____
   4 pieces $1/2'' \times 44''$ rebar
   supply of shop mortar
   masonry tools
   $2'' \times 8''$ form lumber
2. Lay out the project according to plans. Make sure it is square. Check floor or foundation for level. Make adjustments in first course if needed.
3. Dry bond the first course to make sure pilasters will be as on plan. Range the front side when you lay the first course.
4. Dry bond the second course pilasters. When layout is right, finish the first three courses of the project.
5. Lay the pilaster block after you plumb level and range each course on the 4′ sides.
6. Tool when thumbprint hard.
7. Install wood formwork. Make sure it is $3/8''$ above jambs. Use wood shims if needed to secure form. The formwork should stay in place in this practice project because you will *not* pour the lintel with grout.
8. Build a three-course lead at each end of project finishing the 4′ walls to full height.
9. Hang the line on the fifth course to lay the lintel block. Make sure you keep full joints. Do not get mud on the form; it will tip block.
10. Lay the last course of block with the line.
11. When you are done with your project, clean your area and ask your instructor for an evaluation. Do not remove the wood form until you tear down the whole project.

Instructor's Evaluation and Checklist

Student's Name: _____

	Possible Points	Points Off
1. Project set up and properly stocked	5	
2. Safety procedures followed	5	
3. Plumb. Check nine places.	10	
4. Level. Check three places.	10	
5. Height. Check 8" coursing in one place.	10	
6. Bond: Pilasters are properly bonded.	10	
7. Tooling: Holes Smears Tags or fingernails that protrude from face of wall Uniform shape	10	
8. Proper use of trowel and tools. Proper saw cuts.	15	
9. Motion study as described in project plans and good masonry practices. Install four pieces #4 rebar resting on block webs.	15	
10. Neatness. Inside lintel should not have accumulation of dropped mortar.	10	
	100	

TOTAL

FINAL GRADE

_____ **SCORE**

(Required to pass: Beginning 60–80%, Intermediate 80–90%, Advanced 90%)

Instructor's Comments:

Student Competency #38

TASK

Laying block and glass block to the line.

Checking the end face for plumb. (Courtesy of George Moehrle Masonry Co.)

PERFORMANCE OBJECTIVES

- The student shall be able to set up and mark corner pole to build an 8" block wall that is 80" long and 48" high.
- The student shall be able lay glass block to the line.

RELATED LEARNING ACTIVITIES

1. Properly stock your project as described in workbook Competency #2.
2. Observe all safety practices.
3. Read and review Units 16 and 27 of the *Masonry Skills* textbook.
4. Watch the demonstration by the instructor.
5. Review block laying and mortar spreading practices in Competency #5 of this work book.
6. When glass block are laid, a soft expansion joint is used at one jamb and on top of the last course when it is crossed over by other masonry.

162 Student Competency #38

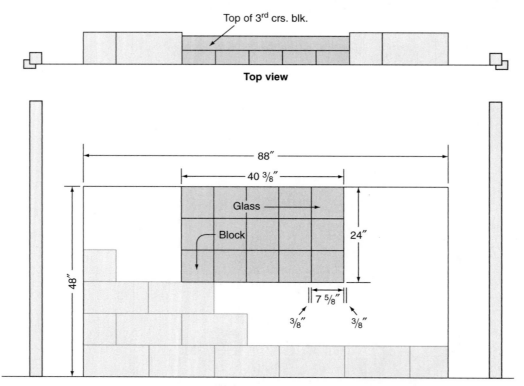

Front view and top view of completed Competency 13.

Checking the inside face for plumb. (Courtesy of George Moehrle Masonry Co.)

Tooling the mortar joints.

PROCEDURE

1. Items needed:
   supply of shop mortar
   8″ block_____
   8″ jamb block_____
   8″ half block_____
   glass block (dry condition)_____
2. Mark the corner poles with the proper coursing. Be careful to check the floor or footing for level and make proper adjustments with the first course of block.
3. Mark out bond on the floor or footing.
4. Hang the mason's line on the corner poles at the marks for the first course.
5. Lay the first course to the line. Remember to keep $1/16''$ away from the line to avoid pushing out of line (range).
6. After laying the first course, check each jamb end for plumb and alignment with the layout marks.
7. Continue this procedure until you have three courses laid.
8. Hang the line for the fourth course and mark out the location for the glass block according to the project plan. Fill the cores at the glass block location with masonry debris or cover with grout screen or hardware cloth.
9. Lay the remaining two courses of block and glass block with the line. Remember to plumb all jambs and the glass block. It is important to check the glass block for plumb because they are being laid in a stacked bond position.
10. Tool all joints when thumbprint hard. The joints in the glass block will take longer to set up. Do not tool them when they are to wet. After joints in glass bock are tooled, carefully wipe excess mortar from the face of the glass block with a rough cloth or old towel. Several passes must be made to get them clean. Do not add water; this will smear them.
11. When you are done with your project, clean your area and ask your instructor for a grade.

Instructor's Evaluation and Checklist

Student's Name: _____

	Possible Points	Points Off
1. Project set up and properly stocked	5	
2. Safety procedures followed	5	
3. Plumb. Check six places.	15	
4. Level. Check top of each course.	10	
5. Height. Check proper course layout and uniform bed joints.	15	
6. Bond. Check for running bond. Check for stack bond on the glass block section.	10	
7. Tooling: Holes Smears Tags or fingernails that protrude out from wall Uniform shape	10	
8. Proper use of trowel and tools	10	
9. Motion study as described in procedure and good masonry practices	10	
10. Neatness. Glass block clean.	10	
	100	

TOTAL

FINAL GRADE

_____ SCORE
(Required to pass: Beginning 60–80%, Intermediate 80–90%, Advanced 90%)

Instructor's Comments:

Student Competency #39

TASK

Laying 4″ block and glass block to the line.

PERFORMANCE OBJECTIVES

- The student shall be able to set up and mark corner pole to build a 4″ block wall that is 80″ long and 48″ high.
- The student shall be able to lay glass block to the line.

RELATED LEARNING ACTIVITIES

1. Properly stock your project as described in workbook Competency #2.
2. Observe all safety practices.
3. Read and review Units 16 and 27 of the *Masonry Skills* textbook.
4. Watch the demonstration by the instructor.
5. Review block laying and mortar spreading practices in Competency #5 of this workbook.

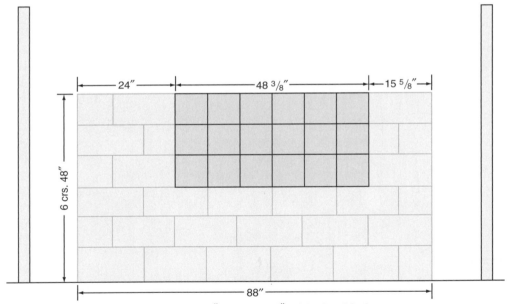

Front view of Competency 39, which is 4″ block and 4″ thick glass block.

PROCEDURE

1. Items needed:
   supply of shop mortar
   4″ block_____
   4″ half block._____
   8″ × 4″ × 8″ glass block (dry condition) _____ (6″ × 4″ × 6″ glass block also works with this project.)
2. Mark the corner poles with the proper coursing. Be careful to check the floor or footing for level and make proper adjustments with the first course of block.
3. Mark out bond on the floor or footing.
4. Hang the mason's line on the corner poles at the marks for the first course.
5. Lay the first course to the line. Remember to keep $1/16″$ away from the line to avoid pushing out of line (range).
6. After laying the first course, check each jamb end for plumb and alignment with the layout marks.

7. Continue this procedure until you have three courses laid.
8. Hang the line for the fourth course and mark out the location for the glass block according to the project plan.
9. Lay the remaining three courses of block and glass block with the line. Remember to plumb all jambs and the glass block. It is important to check the glass block for plumb because they are being laid in a stacked bond position.
10. Tool all joints when thumbprint hard. The joints in the glass block will take longer to set up. Do not tool them when they are too wet. After joints in glass bock are tooled, carefully wipe excess mortar from the face of the glass block with a rough cloth or old towel. Several passes must be made to get them clean. Do not add water; this will smear them.
11. When you are done with your project, clean your area and ask your instructor for a grade.

Instructor's Evaluation and Checklist

Student's Name: _____

	Possible Points	Points Off
1. Project set up and properly stocked	5	
2. Safety procedures followed	5	
3. Plumb. Check six places.	15	
4. Level. Check top of each course.	10	
5. Height. Check proper course layout and uniform bed joints.	15	
6. Bond. Check for running bond. Check for stack bond on the glass block section.	10	
7. Tooling: Holes Smears Tags or fingernails that protrude out from wall Uniform shape	10	
8. Proper use of trowel and tools	10	
9. Motion study as described in procedure and good masonry practices	10	
10. Neatness. Glass block clean.	10	
	100	

TOTAL

FINAL GRADE

_____ SCORE
(Required to pass: Beginning 60–80%, Intermediate 80–90%, Advanced 90%)

Instructor's Comments:

SECTION EIGHT
SCAFFOLDING AND CLEANING MASONRY WORK

Student Competency #40

TASK

Building scaffold

One worker can build the scaffold by putting cross braces on the frame and leaning it on the ground while getting the next end frame set.

PERFORMANCE OBJECTIVE

■ The student shall be able to properly and safely erect a 4′, 4′ 6″, or 5′ mason's scaffold and to install plank and stock materials on it.

RELATED LEARNING ACTIVITIES

1. Read and review Units 28 and 29 of the *Masonry Skills* textbook.
2. Watch the demonstration by the instructor.
3. This scaffold shall be used for Competency test #15 of this workbook.
4. *Caution: The student should only walk on the two lower walk planks. Do not climb or walk on the material planks.*

Have all scaffold parts in place before starting.

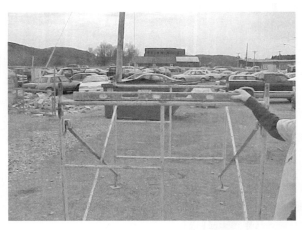
Use the four-foot level to check for level across frame and on plank from one frame to another.

Scaffold set up for competency projects 23 and 29. Notice the mason can stand at a lower level eliminating a lot of bending at the waist.

PROCEDURE

1. Gather and inspect for defects the following supplies:
   2 tubular frame scaffold frames as on project plans
   2 matching cross braces
   4 mud sills at least 10" wide and 24" long
   6 scaffold grade plank 8' long with cleats at each end
   Caution: If building on uneven ground, you will need four adjustable screw legs for the scaffold.
   If any of these components are cracked or damaged in any way, discard them and get proper replacements.
2. Set mud sill in location where scaffold is to be used.
3. Lay the x braces next to mud sills so you can reach them to attach to frames.
4. Stand one frame up with rungs facing wall you are building. Attach x brace to top and bottom connections of scaffold frame. Scaffold can lean on this brace until you set other frame in place.
5. Attach other frame and x brace.
6. Check to see if scaffold frames are 2" from face of wall. Adjust if necessary.
7. Install two walk planks for mason on lower rungs. Install four $9^1/_2$"-wide plank or three 12"-wide plank on top of scaffold frames. Make sure each plank overlaps 6" and has a cleat under each end.
8. Stock scaffold by putting mud board in middle and materials on either side. Leave a space for your tools and level.
9. *If building with screw legs, adjust legs to approximate determined height and insert in bottom of scaffold frames that are lying down.* Lift one frame at a time and attach x brace. Proceed as described above. *Use the four-foot level to plumb and level scaffold. Adjust screw legs as needed.*
10. When you are done with your project, clean your area and ask your instructor for a grade.

Instructor's Evaluation and Checklist

Student's Name: _____

	Possible Points	Points Off
1. Proper components	10	
2. Proper procedure without assistance	10	
3. Mud sills installed	10	
4. Plank overlapped properly	10	
5. Scaffold is plumb and level.	10	
6. Scaffold is correct distance from wall and centered on project.	10	
7. Mud board in middle	10	
8. Material stocked securely. Alternate layers if more than two high.	10	
9. Coupling pins in to prevent mortar buildup in frames	10	
10. Neatness. Area around scaffold clean. No tripping hazards.	10	
	100	

TOTAL

FINAL GRADE

_____ SCORE

(Required to pass: Beginning 60–80%, Intermediate 80–90%, Advanced 90%)

Instructor's Comments:

SECTION NINE
CHIMNEYS AND FIREPLACES

Student Competency #41

TASK

Laying a brick chimney.

PERFORMANCE OBJECTIVE

- The student shall properly lay out and build a brick chimney 64" tall.
- The student shall build in a cleanout door, 6" thimble, and a corbelled brick top.

RELATED LEARNING ACTIVITIES

1. Properly stock your project as described in workbook Competency #2.
2. Observe all safety practices.
3. Read and review Unit 32 in the *Masonry Skills* text book.
4. Review Competency #29, which has a corbelled brick top.
5. Watch the demonstration by the instructor.
6. For the purposes of this lab, no holes will be cut in flue liners or thimbles. Regular mortar shall be used for flues and thimbles instead of code-approved refractory mortar.
7. For the purpose of this lab, wood models of flue liners can be used. Never use wood components in a wood burning fireplace or chimney.
8. Do not build an actual chimney unless you have a skilled mason who understands building code issues with chimneys and fireplaces.
9. Chimneys built in areas prone to earthquakes have vertical and horizontal reinforcing bars built into the masonry and tied to the structure.

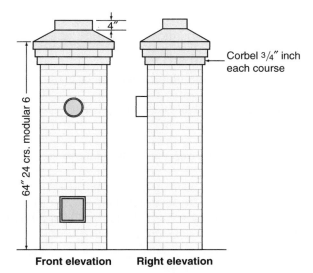

Corbelling the top courses of the chimney add a decorative element and helps the chimney to shed water.

178 Student Competency #41

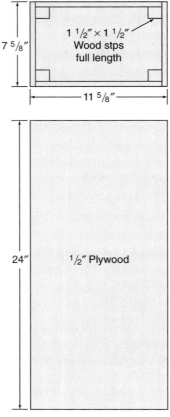

Use plywood or aspenite sheathing and wood strips to build mock-ups of flue liners.

PROCEDURE

1. Items needed: brick_____
   3 7 5/8″ × 11 5/8″ flue liners. Plywood mock-ups of flue liners can be used.
   1 6″ × 8″ thimble
   1 8″ × 8″ cleanout door. (If none is available, build the proper masonry opening for it.)
   supply of shop mortar
   masonry tools
2. Lay out the first course of brick according to project plans. Build the project by laying the brick on the modular #6 scale.
3. Lay the three courses of brick. Mark out 8 1/4″ opening for cleanout door.
4. Lay three courses to top of cleanout opening. Do not install door now. Set a brick on end in the middle of the opening. Use this as a support to lay your brick across to build the chimney. Place some paper bond breakers under and on top of the stiffer mortar that will support the seventh course.
5. Lay the eighth course and then set flue liner in. Continue laying brick, keeping work plumb and level. Check coursing as you lay. Tool joints when thumbprint hard.
6. On the fifteenth course, mark out the location of the thimble and make the proper cuts to go around the thimble. Allow 1/2″ for mortar joints around the thimble.
7. Lay the thimble opening in but do not install the thimble yet. Install this when chimney is done. This prevents bumping it and breaking the "set" of the mortar.
8. Lay up to the seventeenth course and install the next flue liner.
9. Continue laying brick to the twenty-first course.
10. Corbel the last three courses according to project plans.
11. Install a mortar wash with the proper slope.

12. Install thimble. First clean out any dropped mud from opening. Spread bed joint on bottom of opening. Butter edge of thimble that goes against flue. Push thimble in against flue. Press mortar all around thimble with a flat slicker. Repoint when thumbprint hard. Reach into thimble and wipe clean with a damp rag. Fill in any holes.
13. Install cleanout door by spreading bed joint in opening. Butter mud on the outer flange and push it into opening. Supporting door frame with one hand, open door and smooth of bed joint mud. Shut door. Carefully wipe outside edges. Brush lightly. Prop a short piece of wood against it if needed until dry.
14. When you are done with your project, clean your area and ask your instructor for a grade.

Student Competency #41

Instructor's Evaluation and Checklist

Student's Name: _____

	Possible Points	Points Off
1. Project properly set up and stocked.	5	
2. Safety practices followed.	5	
3. Plumb. Check four places.	15	
4. Level. Check four places.	10	
5. Height. Check coursing.	10	
6. Location of cleanout and thimble according to plan.	15	
7. Tooling: 　Holes 　Smears 　Tags or fingernails protruding from wall or joint. 　Uniform shape	10	
8. Proper use of trowel and tools	10	
9. Motion study as described in procedure and good masonry practices	10	
10. Neatness	10	
	100	

TOTAL

FINAL GRADE

_____ SCORE

(Required to pass: Beginning 60–80%, Intermediate 80–90%, Advanced 90%)

Instructor's Comments:

Student Competency #42

TASK

Building a Rumford fireplace.

Rumford thought that fireplaces should be a lot smaller than they were in England in the eighteenth century. They wasted too much heat up the chimney and pulled too much cold air into the room. He also knew from his study of heat that if the fireplace is shallow (ideally one-third as deep as it is wide) and the coving (or sides) are angled a maximum of 135 degrees to the back wall, the fireplace would reflect more heat into the room.

PERFORMANCE OBJECTIVE

- ▪ The student shall properly and safely lay out and build a 36″ Rumford* fireplace with throat, smoke chamber, and brick face.

RELATED LEARNING ACTIVITIES

1. Properly stock your project as described in workbook Competency #2.
2. Observe all safety practices.
3. Read and review Units 33 and 34 of the *Masonry Skills* textbook.
4. Watch the demonstration by the instructor.
5. *The Rumford fireplace is a style of fireplace named after the man who developed it in the 1700s. The angled sides and straight fireback are easy to build and radiate heat. The Rumford is about four times more efficient than a conventional fireplace and burns cleaner. Most people stand the firewood up against the fireback tepee style to get the best combustion and heat.

182 Student Competency #42

6. For the purposes of this competency, shop mortar will be used with the firebrick instead of code-approved refractory mortar. Code-approved refractory mortar allows $1/8''$ joints to be used when laying firebrick. These joints make the firebox and hearth more fire resistant.
7. For the purpose of this lab, no flat Rumford-style damper as supplied by Superior Clay Products or other masonry suppliers shall be used.
8. For the purpose of this lab, a wood model of a manufactured Rumford throat and smoke chamber similar to the actual clay throats and smoke chambers manufactured by Superior Clay Products can be used. Never use wood components in an actual wood burning fireplace or chimney.
9. Do not build an actual wood burning fireplace unless you have the supervision of a skilled mason who understands the building codes for fireplaces and chimneys.

Build the Rumford firebox using standard 9″ firebrick with the 9″ × 4.5″ face showing. Use Heat Stop refractory mortar. Fill any voids and wash the firebox with a sponge and plain water. Rumford fireplaces are usually about as tall as they are wide, but you can vary the height by the number of firebrick courses laid. Before starting the firebox, lay out the first course of firebrick and double-check all the firebox dimensions on the plan.

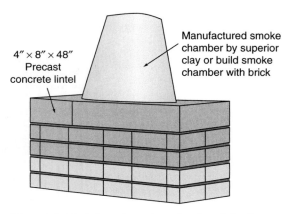

The smoke chamber slopes into the size of the flue liner. When laying the smoke chamber with brick, corbel into the flue liner size.

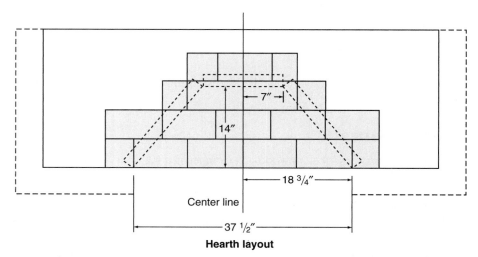

Hearth layout

Layout plan for hearth and first course of firebox. Notice the location of masonry back-up and fireplace face indicated by dotted line.

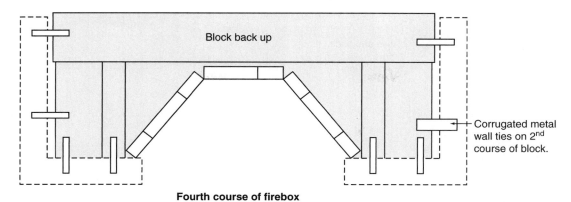

Fourth course of firebox

Fourth course of firebox. Notice the position of block backup and fireplace face that will be laid later.

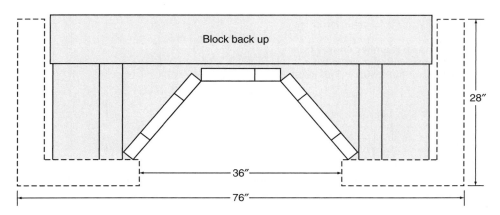

Firebox and masonry backup built to top of firebox. The manufactured throat can be set on and built around.

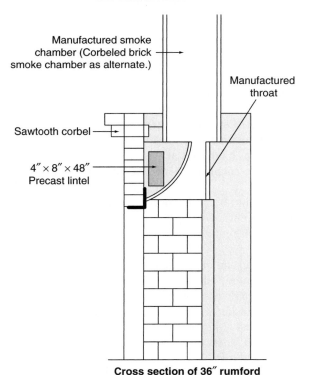

Cross section of 36″ rumford

A corbelled brick throat and smoke may be built as an alternate to manufactured products. Start the throat with a 4″ × 4″ × ³⁄₄″ × 48″ angle iron lintel. Lay brick on lintel and corbel each course. Always keep back plumb.

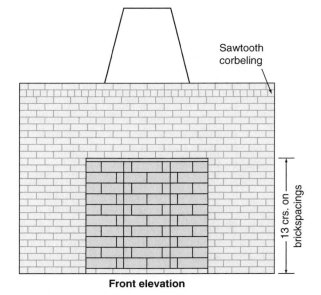

Front elevation

A variety of facing materials may be used other than the brick face specified for this project. Remember to keep top of fireplace opening the proper height.

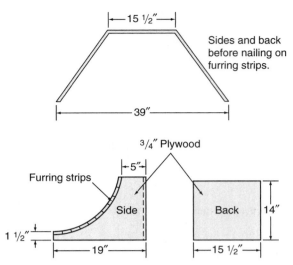

Manufactured throat and smoke chamber units. These units save time and ensure correct design. Wooden mock-up of 36″ Rumford throat built with ³⁄₄″ plywood and furring strips.

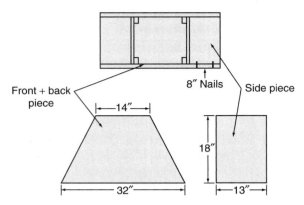

Wooden mock-up of 36″ Rumford smoke chamber. Use ³⁄₄″ plywood and 1¹⁄₂″ × 1¹⁄₂″ wood strips to build it.

PROCEDURE

1. Items needed:
   firebrick_____ (If firebrick is not available, use solid brick with no cores.)
   premade throat section as manufactured by Superior Clay Products (or equal), or wooden mock up of throat for shop use only
   4″ × 8″ × 48″ precast lintel
   premade smoke chamber sections as manufactured by Superior Clay Products (or equal) or wooden mock-up of manufactured smoke chamber. The student can use the option of building the smoke chamber with corbelled brick courses and parging the inside.
   supply of shop mortar
   masonry tools
2. Lay out the first course of the fireplace according to plan. Locate center of front of the project to lay out firebox opening.
3. Dry lay the first course of brick on the front of the fireplace. Start the firebox layout ¹⁄₂″ behind face brick.
4. To lay out a firebox, all measuring will be done from the centerline and front of firebox line. Extend centerline back with framing square. Mark back of firebox at 14″ in from front of firebox. Using the square, measure 7″ each way to establish the fireback, which is 14″. Mark the front of the firebox measuring 18 ³⁄₄″ each way from the centerline. Review layout. If correct, extend lines and lay the hearth firebrick at least 2″ past firebox wall lines on sides and back. They will be laid as shiners according to project plans. Mark the center of several brick lightly. Use these in the center. This will help you stay on bond.
5. Lay the hearth shiner ways as on plan. Prepare a uniform flat bed joint. Butter the brick carefully using ¹⁄₄″ joints. Too much mud will cause excess tapping into place and may push work apart.
6. After hearth is complete, lay out firebox again. Lay the first course of firebrick. Tool joints, sweep hearth, and put a ¹⁄₂″ layer of sand on the hearth.
7. Lay the firebox to the fourth course. Build in block and brick back up and install wall ties to tie fire brick to back up. *Make sure all joints are full!* When laying block backup, put in wall ties to tie face brick later.
8. Lay the last four courses of firebrick, and lay block and brick backup to top of firebox or closest coursing.
9. After firebrick joints have set, use a damp cloth or sponge to float finished joints of firebrick. Wipe at a 45-degree angle across joints to fill in and clean off smears. Rinse and wring out cloth or sponge as needed.
10. Set the premade throat section in mortar on top of the fire box. Build block and brick backup to top of throat. Use the 4″ × 8″ × 48″ precast lintel to span the inward curve of the throat.
11. Set premade smoke chamber on top of masonry being careful to center over throat. If premade smoke chamber is not available, lay a brick smoke chamber as on project plans. Parge the inside smooth. When smoke chamber is done, proceed with brick face.

12. Dry bond first course of brick again on face. Dry bond right through opening to make sure bond works after crossing opening. Make the proper cuts to start project.
13. Lay out coursing to reach the top of opening. The masonry opening should be only $1/2''$ to $1''$ below premade throat. Use the brick spacing ruler to lay out coursing.
14. When laying the face, build a three-course lead on each side and lay in with line. Remember to plumb all jambs.
15. Lay brick to top of opening. Tool joints when thumbprint hard.
16. Set steel lintel in place and continue building the face with a three-course lead at each end and laying to the line. Lay brick until six courses are laid over opening.
17. Lay a brick mantel by corbelling a sawtooth course as on the plans. Then lay a double wythe course as on plans tool the mantel with a flat slicker. *Remember:* Lay the back wythe of brick first to counterbalance the corbelled brick that are laid on the sawtooth brick.
18. When face is done, lay the front course of fire brick in the front of the firebox between the brick face.
19. When the project is done, clean your area and ask your instructor for a grade.

Instructor's Evaluation and Checklist

Student's Name: _____

	Possible Points	Points Off
1. Project set up and properly stocked	10	
2. Safety procedures followed	10	
3. Plumb. Check firebox in three places. Check face in four places. Check block backup in six places.	10	
4. Level. Check hearth in two places. Check lintel. Check mantel top.	10	
5. Height. Check that each side of face is the same coursing. Check that fireplace opening is 34″ to 36″ above hearth.	10	
6. Bond. Check firebrick for bond. Cuts should be uniform. Check face for bond. Cuts should be uniform and in same position on alternating courses. Check that sawtooth corbelling has even layout with no small piece at one end.	10	
7. Tooling: Holes Smears Fingernails and tags that protrude from the face of wall Uniform shape	10	
8. Proper use of trowel and tools	10	
9. Motion study according to project plans and good masonry practices	10	
10. Neatness	10	
	100	

TOTAL

FINAL GRADE

_____ **SCORE**

(Required to pass: Beginning 60–80%, Intermediate 80–90%, Advanced 90%)

Instructor's Comments:

SECTION TEN
ARCHES

Student Competency #43

TASK

Laying an arch in brick veneer.

PERFORMANCE OBJECTIVES

- The student shall be able to properly lay out and build a masonry opening in a 4″ brick veneer spanned by a segmental arch.
- The student shall properly lay out the rowlock course that rings the arch, and cut and lay the brick that meet the arch ring.

RELATED LEARNING ACTIVITIES

1. Properly stock your project as described in workbook Competency #2.
2. Observe all safety practices.
3. Read and review Units 36 and 37 of the *Masonry Skills* textbook.
4. Watch the demonstration by the instructor.
5. The mason can lay a piece of properly sized rope or backer rod on the arch form between the brick in the arch ring. This will eliminate chiseling joist in the arch soffit after form is removed.

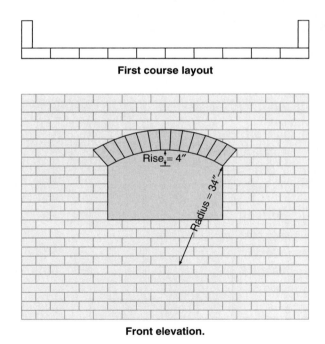

First course layout

Front elevation.

Front elevation.

PROCEDURE

1. Items needed:
   brick_____
   arch centering (form) and shim wedges
   supply of shop mortar
   masonry tools
2. Lay out wall according to project plans. Dry bond the first course of brick. When layout and bond are correct, build a six-course lead at each end. Tool joints when thumbprint hard. The project will be built on the #6 modular scale.
3. After the six course corner leads are finished, lay in the wall.
4. Build six course leads at each end of the project again and lay in the wall.
5. Build six course leads at each end again. Mark out the masonry opening. Lay in the wall to the masonry openings. Plumb each jamb.
6. Set the arch centering in the masonry opening on two 4″ blocks and wedge shims. Make sure it is level and plumb. Dry bond the two skew back bricks and mark for cuts.
7. Using the brick spacing ruler, mark out the position of the rowlock brick that will make the arch ring. Lay the skewback course to the line. Do not forget to leave a closure joint in the middle.
8. Raise the line a course and lay a bat at each end against the skewback brick. The line will help you keep the bats plumb as you lay them ahead of the stretcher courses. Lay bats at each end on the arch form keeping with the layout marks. Be sure the face of the bat is down. Lay in the next course. Determine your cut when you reach the rowlock brick in the arch ring. Lay several more bats in the arch ring. Be sure to stay on the marks.
9. Continue laying to the line and filling in with cuts until you have crossed over arch ring. If coursing was followed and plumb bond maintained, the brickwork will fit evenly across the arch.
10. Continue laying brick until the leads are filled in. Tool joints when thumbprint hard.
11. When your mortar has sufficiently dried (24 hours or more in the practice lab but 7 days or more on a real job), remove arch form and point in the joints under the arch (soffit) with a flat slicker.
12. When you are done with your project, clean your area and ask your instructor for an evaluation.

Instructor's Evaluation and Checklist

Student's Name: _____

	Possible Points	Points Off
1. Project set up and properly stocked	5	
2. Safety procedures followed	5	
3. Plumb. Check seven places.	10	
4. Level. Check two places.	10	
5. Height. Check coursing each end.	10	
6. Bond. Check plumb bond in two places. Check for even layout of headers in arch ring.	15	
7. Tooling: Holes Smears Fingernails and tags that protrude from the surface of wall Uniform-shaped joints that fit around cuts that meet arch ring.	15	
8. Proper use of trowel and tools. Accurate saw cuts appear uniform after tooling is done.	10	
9. Motion study according to project plans and good masonry practices	10	
10. Neatness	10	
	100	

TOTAL

FINAL GRADE

_____ **SCORE**

(Required to pass: Beginning 60–80%, Intermediate 80–90%, Advanced 90%)

Instructor's Comments:

Student Competency #44

TASK

Laying out and constructing a two rowlock semicircular arch.

PERFORMANCE OBJECTIVE

■ The student shall be able to lay out and build a double rowlock arch of two brick wythe construction on a previously built form.

RELATED LEARNING ACTIVITIES

1. Properly stock your project as described in workbook Competency #2.
2. Observe all safety practices.
3. Read and review Units 36 and 37 of the *Masonry Skills* textbook.
4. This is a typical arch used over masonry openings in new or historic buildings. For the wall to span an opening, it does not require the use of steel lintels or beams, which may later rust and deteriorate.

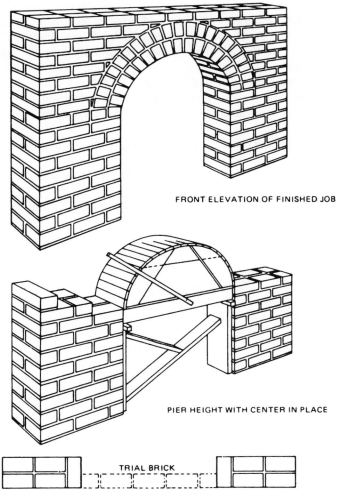

Laying out and constructing a two–rowlock semicircular arch.

193

PROCEDURE

1. Items needed:
   brick_____
   supply of shop mortar
   masonry tools
   semicircular arch form to span a opening $32^{3}/_{8}''$ wide and $7''$ thick.
2. Lay out the project on floor or foundation according to project plans. To lay out, dry bond the brick using $^{3}/_{8}''$ joints. Set the arch center form in place over dry bonded brick to make sure it fits and then remove it.
3. Check floor or foundation for level. If floor is out of level, adjust your first bed joint so top of first course is level.
4. Build brick piers eight courses high to spring point (where arch starts) height as illustrated in project plans. Be careful to keep the piers at the same height (level with each other). Build a two-course lead at each end as on project plan.
5. Set arch form level and plumb with wood supports and shims so that the radial center point is $1''$ above spring point. Install diagonal bracing to stabilize supports as shown on project plans. Attach a radius stick to the radial center point with a small nail.
6. Lay out the brick spacing of rowlock brick that ring the arch. Use a small joint layout such as #2 or #3 on the brick spacing ruler. Remember to have a closure joint. Since the brick spacing ruler will not wrap around the form, lay out the brick spacing marks on a piece of masking tape that is the same circumference as the arch form. After the marks are determined, put the tape on the form and transfer the marks to the face of the form so you can find them as you are laying brick.
7. Hang the line on the leads and lay the rowlock bricks from the edges to the center of the arch. Keep the brickwork filled in to the rowlock brick as they are laid to the intended brick spacing marks. Use the line to keep your rowlock brick plumb with the leads and your stretcher brick level. Insert $^{1}/_{4}''$ rope or backer rod in the face of the bed joint facing the form. This will make it easy to clean out joints after you remove the form and point in the soffit (bottom) of the arch.
8. The face of the rowlock brick needs to be laid flat against the form as you work your way around the curve of the form. This will prevent the jagged edge look after the form is removed.
9. Keep the leads at each end built ahead of your work so you have a line as a guide to keep the arch ring plumb and in range.
10. As you build the arch ring, cut the bricks as needed making a uniform joint.
11. Continue building the project as on the plans.
12. When the project has cured (at least 24 or more hours and on real jobs at least 7 or more days), remove arch form and point in soffit of arch.
13. When you are finished with your project, clean your area and ask your instructor for an evaluation.

Instructor's Evaluation and Checklist

Student's Name: _____

	Possible Points	Points Off	
1. Project set up and properly stocked	5		
2. Safety procedures followed	5		
3. Plumb. Check seven places.	10		
4. Level. Check two places.	10		
5. Height. Check coursing in one place.	10		
6. Bond. Check pattern bond of project. Check for uniform brick spacing of rowlock brick around arch ring.	15		
7. Tooling: Holes Smears Fingernails or tags that protrude from the face of wall Uniform joints formed around cuts that meet rowlock brick	15		
8. Proper use of trowel and tools. Accurate saw cuts that appear uniform after tooling is done.	10		
9. Motion study according to project plans and good masonry practices. Arch form set in a level secure position at proper height.	10		
10. Neatness	10		
	100		TOTAL
			FINAL GRADE

_____ **SCORE**

(Required to pass: Beginning 60–80%, Intermediate 80–90%, Advanced 90%)

Instructor's Comments:

SECTION ELEVEN
CONCRETE AND STEPS

Student Competency #45

TASK

Using concrete hand tools.

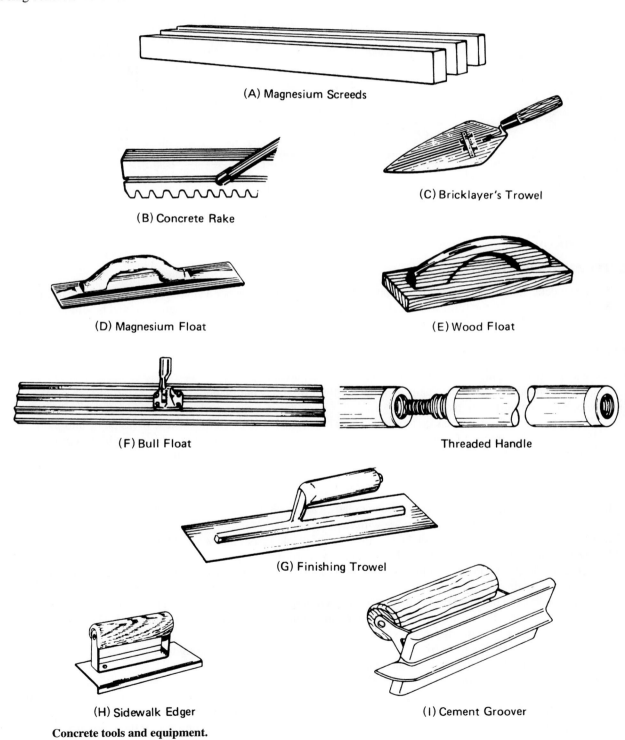

Concrete tools and equipment.

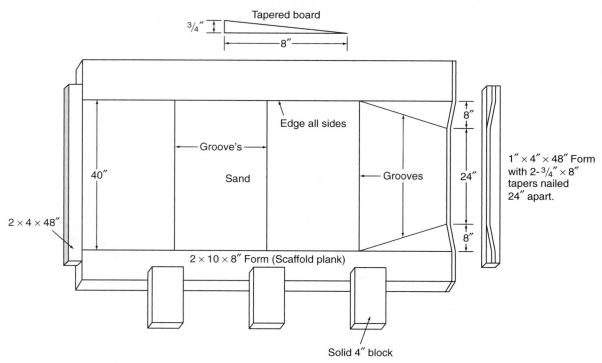

Formwork layout.

PERFORMANCE OBJECTIVE

- The student shall be able to safely and properly lay out and put together a simple concrete sidewalk form.
- The student will fill the form with sand and use the screed, float, trowel, edger, groover, and broom to build a sidewalk mock-up with damp, loose sand.

RELATED LEARNING ACTIVITIES

1. Read and review Units 39 and 40 of the *Masonry Skills* textbook.
2. Check all plank for protruding nails or splinters. Remove if any are found.
3. Watch the demonstration by the instructor.
4. The temporary forms shall be weighted down with block to hold their position instead of fastening into shop floors.
5. If your school or training facility has rubber concrete stamps available, they may be used with this project. Stamping concrete takes careful planning and layout to achieve a uniform pattern.

PROCEDURE

1. Lay out project on floor according to project plans. Use two 8' scaffold plank laid flat as forms. Position along the layout lines. Be sure ends are square with each other.
2. Weight the plank down with block laid on the edge of plank. Install the two end forms the same way.
3. Use the wheelbarrow to bring damp, loose sand to your form. Dump it in and rake to just above form level.
4. Start at the level end and screed the sand with a back-and-forth motion as you move forward. When you are 2' from the ramp end, stop and screed across the end of the form and along the screeded sand. This curb cut ramp should be maintained during the finishing operations. Fill in any low spots and rescreed.
5. Float the surface with the mag float. Mark out control joints every 2' starting from one end. Mark both forms so you can line up a straight edge to put grooves in.
6. Tool in the grooves. Keep the front of the groover up a little as you pass through the sand. Make a distinct groove. Any residual lines can be floated out later.
7. Use the edger to edge along all the forms. Keep the front of the edger up slightly as you move it through the sand.
8. Use the float and then the finishing trowel to trowel down the surface of the sand to a smooth uniform finish. If residual lines from the groover or edger are still there, float area again to achieve a flat uniform surface.
9. Use a soft bristle push broom to pull across the surface to achieve a broom finish. Lightly retool grooves and edge, leaving uniform lines next to the broomed surface.
10. When you are done with your project, clean your area and ask your instructor for an evaluation.

Instructor's Evaluation and Checklist

Student's Name: _____

	Possible Points	Points Off
1. Proper layout of forms and spacing of grooves/control joints	20	
2. Uniform tool marks	20	
3. Uniform finish	20	
4. Proper use of tools	20	
5. Neatness	20	
	100	

TOTAL

FINAL GRADE

_____ SCORE

(Required to pass: Beginning 60–80%, Intermediate 80–90%, Advanced 90%)

Instructor's Comments:

Student Competency #46

TASK

Laying brick pavers in sand.

Basketweave paving in a courtyard.

PERFORMANCE OBJECTIVES

- The student shall be able to safely and properly lay brick pavers in a sand bed.
- The student shall be able to properly lay out the running bond, basketweave, and herringbone patterns.

RELATED LEARNING ACTIVITIES

1. Review Competency #40 of this workbook because a screed and float or finishing trowel will be needed to level sand.

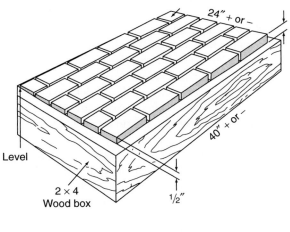

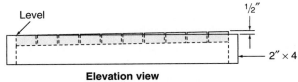

Laying brick paving in a running bond.

Herringbone brick walk.

Student Competency #46

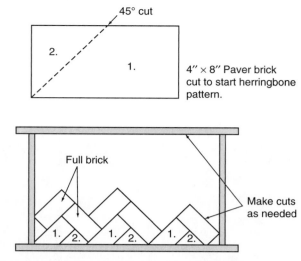

Keep the cuts uniform and use a mason's line or straightedge to keep the joints straight.

PROCEDURE

1. Items needed: paver brick (clay or concrete) (4″ × 8′ or $3^3/_4″ × 7^5/_8″$ solids)
   mason tools
   sand (Please keep free of brick chips when you clean up project.)
2. Build wood form made from 2″ × 4″ big enough to lay five paver bricks long (40″ approximate) and six paver brick wide (24″ approximate). Dry bond the brick to make the form fit the brick with $1/_4″$ of play on two sides. This will accommodate any variation in brick size so you do not have to force brick in the form. The corners of this form should be squared.
3. Fill the form with enough sand to lay the brick as on project plan. The sand should be screeded and floated to smooth uniform finish.
4. Lay the four corner bricks to the proper height. Dry bond the first and lowest course of brick. If bond layout is correct, lay the course and tap the brick lightly into place with trowel handle or rubber mallet. Use a four-foot level or string line to straight edge and level. If brick are low, pick up, sprinkle a little sand under, and lay again.
5. Lay the next course. Be careful to place the brick directly next to the course in place. *Do not slide the brick against the brick in place. You will push sand in between them.*
6. Procedure for running bond: Cut two halves. Mark center of brick by laying another paver brick over it the opposite way. Keep end flush and draw a pencil line. Cut directly on the line. This works with paver brick and fire brick only! Lay brick in running bond using halves at edges or border as need. Use a line or straightedge to keep courses even. Slight variation in brick will cause a wavy course unless a line is followed.
7. Procedure for basketweave: Prepare sand bed. Dry bond or layout to each corner. Set a pair of bricks in each corner at the proper bond and height. Lay bricks two at a time in course before continuing with next course. Always use a line or straightedge. On long runs, use both or two lines each way to maintain the bond in a straight line.
8. Procedure for herringbone: Prepare sand bed. Dry bond and lay out bond to each corner using 45-degree angle cuts. To make proper cuts, do the following: Mark the center of a paver brick. Then from the center edge draw a diagonal line to the corner (45 degrees). Cut the brick right on the line. You will use both pieces. Start in the lower left-hand corner and set the diagonal cut of the full brick along the edge or form or border. The point of the brick should be in the corner. Lay the full brick with the end flush with the upper right-hand corner of starter brick and the bottom corner against the form or border. Continue this procedure across the near side of the form or border until you reach the right hand side. Make the original cut and fit in pieces as needed. The cuts will remain the same as long as you maintain a true 45-degree angle and straight pattern of brick. A border of full brick around the herringbone pattern helps hold the smaller cuts in place. Run a string at a 45-degree angle from the starter course to keep the pattern straight.
9. When you are done with one of the projects, sweep sand into the joints and then sweep clean. Clean your area and ask your instructor for an evaluation.

Instructor's Evaluation and Checklist

Student's Name: _____

	Possible Points	Points Off
1. Project properly set up and stocked	5	
2. Form is uniform dimension and square.	10	
3. Levels. Check two places.	10	
4. Flat paving. Check four places with straightedge. There should be no low or high bricks.	10	
5. Coursing. Check each course for straight alignment.	10	

6. Bond. Proper bond and even cuts
    Running bond
    Basket weave
    Herringbone
	10
	10
	10	

7. Motion study according to project plans and good masonry practices.	10	
8. Proper use of saw or hammer and brick set.	10	
9. Neatness.	5	
	100	

TOTAL

FINAL GRADE

_____ SCORE

(Required to pass: Beginning 60–80%, Intermediate 80–90%, Advanced 90%)

Instructor's Comments:

Student Competency #47

TASK

Mixing concrete and pouring step treads.

PERFORMANCE OBJECTIVES

- The student shall be able to properly form a 13" × 27" × $2^{1}/_{4}$" thick concrete step tread.
- The student shall be able to properly mix, pour, and finish a concrete step tread.
- The student shall understand the difference between a concrete mixer and a mortar mixer.

RELATED LEARNING ACTIVITIES

1. Read and review Units 39 and 40 of the *Masonry Skills* textbook.
2. Review Competency #41 of this workbook.
3. Check all form boards for protruding nails or splinters.
4. Use proper safety procedures when working with wood-cutting hand tools.
5. Use these treads for Competency #48, "Building brick steps with precast treads." The size and thickness of this concrete project are similar to quarried flagstone treads or precast treads.
6. Use caution when mixing concrete to avoid splashing on skin or in eyes.

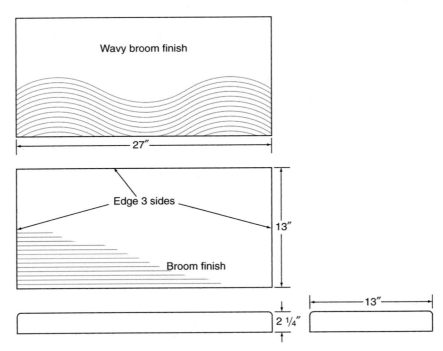

Dimensions of finish concrete treads.

PROCEDURE

1. Gather the proper supplies of:
   furring strips (or $3/4''$ boards ripped to $2^1/_4''$)
   8 d nails
   brick to hold forms
   sand
   gravel
   portland cement
   24" piece of block joint reinforcing or 10 gauge concrete wire.
2. Lay out project on floor. Be sure to square it and check that the dimensions are correct.
3. Cut the form boards from a standard furring strip purchased at a lumber yard or have the instructor rip wood to the proper form height.
4. Nail the ends together as on the project plan. Use the best sides and corner facing the concrete.
5. Place the form on a piece of polyethylene sheeting or a garbage bag to prevent the concrete from bonding to the floor. Dampen inside of forms with soapy water. Place bricks snugly around form to keep it from pushing when concrete is poured in.
6. Mix the concrete in a wheelbarrow by hand or in an approved concrete mixer (not a mortar mixer). Add 3 large coffee cans sand, 1 large coffee can portland cement, and 2 large coffee cans of $1/_4''$ to $1/_2''$ clean gravel. Mix thoroughly until all ingredients are blended. Add a small amount of water and mix thoroughly with dry mix. Continue adding small amounts of water and mixing until it is the thickness of stiff oatmeal. (Water should not separate from mix materials.)
7. Shovel mix into form $1^1/_2''$ thick. Place reinforcing wire in form. Keep the wire reinforcement at least 1" away from edge of form. Shovel in mix to top of form.
8. Screed the concrete with a straight furring strip or hand float.
9. Float surface to uniform flatness that is even with top of form. Lightly use edging tool on all sides. (The concrete mix is too wet to achieve a flat, smooth-troweled surface at this point.)
10. If there is concrete bleed water, allow it to evaporate or absorb bleed water with paper towels and discard so cement residue does not get on your skin or other students.
11. Float again after bleed water is off. Use concrete edger on all sides and float again to a uniform flat surface.
12. After the surface is thumbprint hard, trowel smooth and lightly broom the surface in a uniform pattern of your choice.
13. Alternate surface finish. After floating and the surface is tacky, you can make a brick or stone impression in the concrete, Place 2 mil polyethelene sheeting over surface. Use a narrow, concave striking tool or carpenters pencil and carefully tool in joints to resemble brick or stone.
14. If the imprinted pattern is not satisfactory, refloat surface and try again or use the broom finish.
15. When finished, cover carefully with polyethylene sheeting or large garbage bag. Allow concrete to cure for 7 days and then remove from form. Edges may be rubbed with a piece of concrete brick or a carburundum rubbing stone. Use a little water to work up a bit of concrete paste to fill pinholes. Clean the forms so somebody else can use them.
16. When you are done with your project, clean your area and ask your instructor for an evaluation.

Instructor's Evaluation and Checklist

Student's Name: _____

	Possible Points	Points Off
1. Proper lay out of form and prepare for pour	20	
2. Cast concrete has uniform edges and flat surface.	20	
3. Proper use of tools and curing techniques	20	
4. Cast concrete has uniform appearing finish on top.	20	
5. Neatness	20	
	100	

TOTAL

FINAL GRADE

_____ SCORE

(Required to pass: Beginning 60–80%, Intermediate 80–90%, Advanced 90%)

Instructor's Comments:

Student Competency #48

TASK

Laying brick steps with precast concrete treads.

PERFORMANCE OBJECTIVE

- The student shall properly lay out and build brick steps three risers high.

RELATED LEARNING ACTIVITIES

1. Properly stock your project as described in workbook Competency #2.
2. Observe all safety practices.
3. Use the precast treads from Competency #47 of this workbook.
4. Steps of this design can use various types of flagstone treads or precast concrete treads.
5. Steps such as this should always be built on a concrete slab or block foundation that is protected from frost or settling.

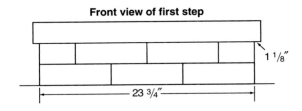

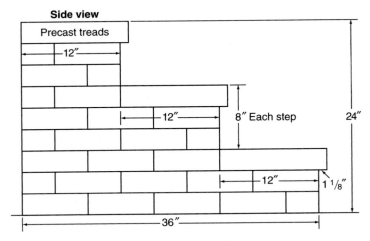

Build the steps before setting precast treads.

PROCEDURE

1. Items needed:
   supply of shop mortar
   masonry tools
   brick
   three precast concrete treads measuring 13″ × 27″
2. Lay out project on floor according to project plans. If possible, build the project against an existing wall or project wall. Be sure the floor is level. If it is not, adjust the bed joint under the first course to make the first course level.
3. Dry bond the first course and mark proper head joint spacing.
4. Lay the first two courses. Tool joints when thumbprint hard.
5. Lay the inside courses to provide support for the tread and the next riser.
6. Lay the next three courses. Be sure you start on bond.
7. Lay the inside courses to provide support for the second tread and third riser.
8. Lay the last three courses of brick and the inside courses of brick to provide support for the third tread. Tool joints when thumbprint hard.
9. Proceed with setting the treads. Set the top tread on first to avoid dropping mortar on lower treads. Spread a uniform bed joint on all sides. Set tread on carefully. Level both ways. Adjust the tread to slope $1/8''$ toward the front of the steps. Check to see if the overhang is the same on the sides and front. Set the other treads with the same procedure.
10. When you are done with your project, clean your area and ask your instructor for an evaluation.

Instructor's Evaluation and Checklist

Student's Name: _____

	Possible Points	Points Off
1. Project set up and properly stocked	5	
2. Safety procedures followed	5	
3. Plumb. Check six places.	10	
4. Level. Check each tread for level and slope.	15	
5. Height. Check each tread for height.	15	
6. Bond. Check for proper bond pattern.	10	

7. Tooling:
    Holes
    Smears
    Fingernails or tags that protrude from face of wall
    Uniform appearance
	10	

8. Proper use of trowel and tools	10	
9. Motion study according to project plans and good masonry practices	10	
10. Neatness	10	
	100	

TOTAL

FINAL GRADE

_____ SCORE

(Required to pass: Beginning 60–80%, Intermediate 80–90%, Advanced 90%)

Instructor's Comments:

Student Competency #49

TASK

Laying brick steps with rowlock treads.

PERFORMANCE OBJECTIVE

- The student shall be able to properly lay out and build brick steps three risers high using brick rowlocks for treads.

RELATED LEARNING ACTIVITIES

1. Properly stock your project as described in workbook Competency #2.
2. Observe all safety practices.
3. Steps such as these should always be built on a concrete slab or block foundation that is protected from frost or settling.

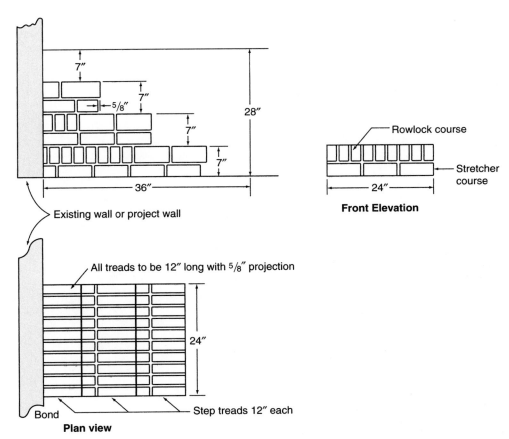

Layout for brick steps. Mark a horizontal line on an existing wall 28″ above finish floor. Build to 7″ below this line.

PROCEDURE

1. Items needed:
   supply of shop mortar
   masonry tools
   brick without cores (Cored brick can be used, but end brick with cores will detract from appearance.)
2. Lay out project on floor according to project plans. If possible, build the project against an existing wall or project wall. Be sure the floor is level. If it is not, adjust the mortar joint under the first course to make it level.
3. Dry bond the first course before starting.
4. Lay the first course and then the first tread, which is two rowlock courses. Slope the rowlock course $1/8''$ toward the front of the steps. Tool when thumbprint hard. Sprinkle a light layer of sand on the first tread before you lay the next courses.
5. Proceed building the steps in this manner according to project plans. Tool the joist when thumbprint hard.
6. When you are done with your project, clean your area and ask your instructor for an evaluation.

Instructor's Evaluation and Checklist

Student's Name: _____

	Possible Points	Points Off
1. Project set up and properly stocked	5	
2. Safety procedures followed	5	
3. Plumb. Check six places.	10	
4. Level. Check each tread for level and $1/8"$ slope to front.	15	
5. Height. Check for proper rise at each step.	15	
6. Bond. Check for proper bond pattern.	10	
7. Tooling: Holes Smears Fingernails or tags that protrude from the face of wall Uniform shape	10	
8. Use of trowel and proper tool techniques	10	
9. Motion study according to project plans and good masonry techniques	10	
10. Neatness	10	
	100	

TOTAL

FINAL GRADE

_____ **SCORE**

(Required to pass: Beginning 60–80%, Intermediate 80–90%, Advanced 90%)

Instructor's Comments: